POWER AND CONTROL IN THE IMPERIAL VALLEY

Connecting the Greater West Series
Sterling Evans, Series Editor

POWER AND CONTROL IN THE IMPERIAL VALLEY

Nature, Agribusiness, and Workers on the California Borderland, 1900–1940

BENNY J. ANDRÉS JR.

WITH A FOREWORD BY STERLING EVANS

Texas A&M University Press
College Station, Texas

Second printing, 2016

Connecting the Greater West Series

Manufactured in the United States of America
This paper meets the requirements of
ANSI/NISO, Z39.48-1992
(Permanence of Paper).
Binding materials have been
chosen for durability.

Library of Congress Cataloging-in-Publication Data

Andrés, Benny J., Jr., author.
Power and control in the Imperial Valley: nature, agribusiness,
and workers on the California borderland, 1900–1940 /
Benny J. Andrés Jr.; with a foreword by Sterling Evans.—First edition.
pages cm—(Connecting the greater west series)
Includes bibliographical references and index.
ISBN 978-1-62349-197-0 (cloth: alk. paper)—ISBN 978-1-62349-219-9 (ebook)
ISBN 978-1-62349-463-6 (paper)
1. Imperial Valley (Calif. and Mexico)—Economic conditions. 2. Imperial Valley
(Calif. and Mexico)—Social conditions. 3. Imperial Valley (Calif. and Mexico)—
Race relations. 4. Irrigation farming—Economic aspects—Imperial Valley
(Calif. and Mexico) I. Title. II. Series: Connecting the greater west series.
F868.I2A54 2015
979.4'99—dc23

2014019851

For my children Abram, Elena, Denisia, and Sevilla

Contents

List of Illustrations ix

Foreword by Sterling Evans xi

Preface xiii

Introduction 1

Chapter 1: "Subjugating" the Colorado River 11

Chapter 2: Colonizing the Desert 40

Chapter 3: Racial Agribusiness 68

Chapter 4: Racial Labor 98

Chapter 5: The Gospel of Labor Rebellion 127

Conclusion 159

Afterword 163

Notes 177

Bibliography 205

Index 219

Illustrations

"Desert Conquorers"(*sic*) 2

Lower Colorado River Basin, 1905 5

J. Edgar Ross tarred and feathered, October 1918 7

Steamboat on the Colorado River near international border, ca. 1906 15

Mexican Colorado River Commission on steamboat in Alamo Canal, ca. 1906 23

All-American Canal dedication, 12 October 1940 37

Mules smoothing the edge of the New River Canyon, ca. 1910 44

Governor Colonel Esteban Cantú and military officers and dignitaries in the Mexicali military fortress, 1919 53

"Ruth" dredger removing silt from canal 61

Imperial County Farm Bureau member transporting bags of grasshopper poison 63

S. A. Gerrard Company cantaloupe packing shed 74

The Sanbonmatsu family, ca. 1926 80

Multiethnic laborers harvesting carrots in open-air factory 107

Agricultural Workers' Industrial League defendants outside County Courthouse, 1930 142

Ejidos planned for the Mexicali Valley, 1938 155

Foreword

Benny Andrés has produced a very well researched and written manuscript that will find itself in good company with the already robust, but continually growing, literature on the history of the US-Mexico borderlands. Here we have a case study of one valley—an important valley in the history of US and Mexican irrigation and one that straddles the border with both similar and different aspects on each side of the line. This is an important and needed work, especially as it deals with so many historical issues within that geographical setting (labor, environment, agriculture, business, government policies) to tell a complete story.

For all of these reasons, *Power and Control in the Imperial Valley: Nature, Agribusiness, and Workers on the California Borderland, 1900–1940* makes an essential addition to the Connecting the Greater West Series. Number three in this new series, the book adds to the ongoing works that explore the changing and growing way historians and others are starting to view the North American West—a region that includes not only the American West, but also northern Mexico, western Canada, the Pacific Rim, and the borderlands between these regions.

Power and Control in the Imperial Valley represents some very sound and competent scholarship. What sets it apart from other similar works is its detailed attention to one specific region and how Andrés connects the valley contextually to US and Mexican policies in general, spending equal time on both sides of the border. There is a multifaceted set of objectives in this book, but they all revolve around the idea of how settlers in the Imperial Valley "sought to impose order on the environment and eliminate social and economic challenges." In many ways, their efforts were to ensure a "conquest of the environment"—working with an understanding that "Mother Nature's grip on the desert wilderness had to be loosened for agribusiness and civilization to flourish."

By using a wide assortment of archival materials (on both sides of the border), newspaper accounts, business records, interviews, and a thorough run on the secondary literature, Andrés pieces together a compelling story that needs to be told. While the story is about one valley, it has much broader implications for borderlands, transnational, agricultural, and environmental history. Note especially Andrés's discussion of the development of the border cities of Mexicali/Calexico for just one of the many fine examples of borderlands and transnational history. Additionally, he appropriately contextualizes these facets with what else was going on in the rest of the United States and Mexico at the time.

What really makes this book unique is the way it ties together so many different dimensions to tell a complete story of the Imperial Valley. It is first and foremost a borderlands history, but it is also certainly an environmental history, an agricultural history, a business history, a labor history, a race and ethnicity history, a political history, a transnational history, and, in many ways, a social history. It also easily falls in the rubrics of the history of the American West and California history. Thus, in the pages that follow, you will discover a remarkably large story about a relatively small area of the Greater West.

Sterling Evans
University of Oklahoma

Preface

I recall floating down the Colorado River on an inner tube near Yuma, Arizona, one summer day when I was a teenager. My friends and I frolicked in the water, unaware that this warm, muddy river was the lifeblood of the arid Southwest. Although I was born and raised in the Imperial Valley, I knew nothing about the history of the region. It never dawned on me how miles and miles of crops could grow in a place where it rarely rained. I never wondered why there were no farmhouses in the countryside. Even though my extended family worked in agriculture, and my father was a forklift driver for produce packinghouses, I lacked understanding of the relationship between farmers and laborers. My knowledge of the history of the US-Mexico borderlands, I am embarrassed to admit, was even more limited. These recollections explain my motivation for writing this history of the California–Baja California borderlands. What follows is a summation of the evolution of this book, as well as a statement of appreciation for the people and institutions that helped me on my journey.

After a stint in the army, I moved to San Diego to attend college. My introduction to borderland's studies began during an undergraduate capstone class at San Diego State University. In response to a term paper I submitted on the exploits of Texas social protester Catarino Garza, my professor, the late Paul Vanderwood, took me to lunch and urged me to attend graduate school. He looked on with a bemused smile when I naively replied, "What is that?" After dispatching me to research graduate schools, Paul convinced me to give it a try and wrote a letter of recommendation for me to attend the University of New Mexico. I am indebted to Paul for launching my career.

UNM had a superb collection of scholars of the American West and Mexico who encouraged me to write a master's thesis and doctoral dissertation on the history of the Imperial Valley. I express my deepest gratitude to my dissertation advisors Betsy Jameson, Ramón Gutiérrez, Linda Hall, and the late Ferenc Szasz for supporting this project in its early stages. After completing graduate coursework, I taught at Imperial Valley College and wrote my dissertation. While living in Brawley I stumbled on the entire run of the *Brawley News* in the local library. I thank Marjo Mello, library director, for granting me access to the newspaper. Assistant librarian Yolanda Gasca lifted my spirits as I slogged along for a year typing a thousand pages of notes. A perusal of the endnotes reveals the importance of the *Brawley News* for supporting my thesis.

I am grateful for the financial support that made this book possible. I thank the University of New Mexico history department for a Kennedy dissertation research grant. Tobías Durán, the director of the UNM Center for Regional Studies, blessed me with a year of financial support. Both the UNM graduate students' association and the Huntington Library supplied two small summer travel grants.

After writing the dissertation, I moved my family across the country to teach at the University of North Carolina, Charlotte. Immediately, I began transforming my case study of the Imperial Valley into a regional history of the borderlands encompassing both halves of the Imperial-Mexicali Valley. As I waded through countless books and articles, I despaired over the means to pay for research trips to California. Fortunately, my colleagues at the University of North Carolina, Charlotte, supported this project with two faculty research grants, which culminated in two journal articles. Nancy Gutiérrez, the dean of the College of Liberal Arts and Sciences at the University of North Carolina, Charlotte, announced her unflagging support with a Schley R. Lyons Junior Faculty Development Award for a semester leave of absence to conduct research and to revise the manuscript. My late father, Benny Sr., underwrote the rest of my research travel.

Archival sources are the lifeblood of interpretive scholarship. Here, I thank the following staff and administrators for granting access to their archival collections: the Bancroft Library at the University of California, Berkeley; the Hoover Library at Stanford University; the Huntington Library; Bill Hendricks, Paul Wormser, and Jill Thrasher at the Sherman Library; Carrie Marsh and Lisa Crane at the Honnold/Mudd Library at the Claremont Colleges; Melissa Conway and Sarah Allison at the University of California, Riverside; the University of California, San Diego special collections; and the staff at the National Archive at Laguna Niguel. I also thank the University of California, Irvine, and the San Diego Public Library for sending me documents from their special collections.

At the tail end of my research, Linda Vida, Jessica Davila Greene, and Tiffany Chow at the Water Resource Collection and Archive at the University of California, Riverside, provided a congenial environment to conduct research. Thank you for granting me permission to include photographs from the water collection. Lynn Housouer and Mike Anderholt at the Pioneers Museum in Imperial County deserve sainthood for patiently filling my numerous requests. I selected the bulk of the pictures from thousands of digitalized images in the Pioneers Museum photograph collection. Interlibrary staff at the University of New Mexico, Imperial Valley College, San Diego

State University, Calexico, and the University of North Carolina, Charlotte, tracked down obscure books and articles. In particular, I acknowledge research librarians Carlos Zazueta, Lois Stickell, and Amanda Binder for their assistance. Ann Davis went the extra mile to track down thirteen years of the *Calexico Chronicle* on microfilm.

As I embarked on the daunting task of writing this book, I launched into a crash course on the twentieth-century history of Mexico and the Mexicali Valley. David García, Tim Asamen, Saraí and Alicia Carrillo, Susana Hernández and her siblings Karla, Guillermina, and Luis helped in innumerable ways. Family and friends assisted me at different stages of my research. I thank Mary Helen and David Hoskins, Denisia and Jim Chatfield, Eric and Heather Andrés, Celi and Anthony Andrés, Tom and Lisa Pauzar, and Denise Pan and Chris Arne for lodging and hospitality. Norma Aguilar introduced me to "El Asalto a las Tierras." Dudley Gardner, a fellow traveler from graduate school, urged me onward. John Martin read the manuscript and passed along suggestions. John Bender lifted my spirits. Mary Ann Irwin, José Alamillo, and Cheryl Hicks imparted sage advice on how to get a book published.

I express my appreciation to Sterling Evans for including this book in his Connecting the Greater West series. I thank Mary Lenn Dixon and Katie Duelm at Texas A&M University Press, for shepherding the manuscript into print. Dawn Hall merits special recognition for skillfully editing the manuscript. It is impossible for me to fully express my gratitude to my wife Tamara, who patiently listened as I revised half-baked ideas into defensible arguments. She read every word and gently suggested changes. To my children Abram, Elena, Denisia, and Sevilla, who watched me labor on this project for all these years, this book is dedicated to you.

POWER AND CONTROL IN THE IMPERIAL VALLEY

Introduction

In December 1939, the Brawley Elks Lodge held a memorial service for deceased members of the organization. Reverend M. Palmerlee, pastor of the First Christian Church, eulogized Imperial Valley's "pioneers," remarking: "Many of these, your brethren, were the men who helped transform a virgin waste into a fertile paradise." Palmerlee reminded the audience of the difficulty of irrigated desert farming. "I say without any conscientious misgivings that they fought a good fight, they finished a well run course, they handed down to you the faith which made them heroes and conquerors."[1] This passage suggests how ideologies of environmental conquest played an integral role in the settlement of the Imperial Valley, an arid agricultural zone along the California–Baja California borderland.[2] The domination of the environment is a central historical theme of this region. The subjugation of nature originated with the struggle to control Colorado River flooding. Victory over the river established a mindset among settlers that the environment could be transformed to maximize agricultural production.[3]

In speeches like the above-referenced eulogy and in the widely read booster tracts of Imperial Valley colonization, writers used rhetorical language of a potential Eden in the grip of the desert's unforgiving grasp. Mother Earth's hold on wilderness had to be loosened for agribusiness and for civilization to flourish. In the lexicon of conquest, colonizers employed metaphors and analogies—images rooted in cultural and religious memory—to mark the perilous landscape. Writers described the Colorado Desert as "waste," "barren," "worthless," "savage," and "sterile." These gendered and racialized descriptions of nature were common at the turn of the nineteenth century. Americans alternated between hating and fearing the desert and the Colorado River, while extolling its agricultural potential. Commentators alluded to the feminine beauty and wildness of the landscape and its yearning to succumb to man's dominance.

George Wharton James, a sensitive chronicler of the desert environment, oscillated between revering its arid beauty and power and a strong impulse to subjugate it. In *The Wonders of the Colorado Desert*, he wrote: "The desert fills one with this passion to know, and in knowing to conquer. Its cry, like that of Life, is irresistible." In the end, he could not reconcile his love and fear of barren land. "So the desert stands before us as a place of contradictions, a place

"Desert Conquorers"(*sic*): Colonizers digging the Eastside High Line Canal, 1913. Courtesy P2000_023_001, Imperial County Historical Society, Pioneers Museum, Imperial, California.

of sharply defined contrasts." After professing his love for the desert, he asked, "Is it a vestal virgin, capable of all the rich beauty and glorious fullness of a completed life, or is it an old woman, worn out, hideous, and ugly, after a life the sweetness and beauty of which have been burned out by evil passions?" The former minister alternated between lusty prose and spiritual meditations: "How earth and heaven meet on the desert! In the far-away blue on the remote horizon the yearning heaven reaches down to the emulous earth, and they become one. So God reaches down and joins himself to man, when the creature, in his noble aspirations, reaches upward to the Creator. Thus man becomes as God, able to do all things, as Christ said he should." God commanded: "The sand-hills, the unfruitful soil, the great valley stretches, must be turned into gardens of Eden in which one may walk and talk with God and commune with angelic and human companions."[4]

Other visitors to the Colorado Desert viewed the landscape through a gendered prism. After traveling by horseback across the area in 1919, J. Smeaton Chase, a critic of modernity, noted that the desert was Janus-faced: beautiful but deadly. "Over a region of mesas of baked clay broken by gullies that forced us into tedious detours, we slowly" crossed the landscape. "If the gaunt and desolate has a degree of perfection, here surely it is reached. I do not see how Sahara, Gobi, or Arabia could improve on this for rigid nakedness and sterility. One here sees Mother Earth scalped, flayed, and stripped to the skeleton. Yet there is a strange beauty in it all." As Chase trudged toward Mexico, the landscape abruptly changed to "intense dreariness, a plain of dust with a scattering of desert plants more than usually wretched and unkempt. A few ocotillos alone broke the torrid stillness with a skeleton dance on the quivering air." After three weeks wandering under the burning sun, he rode into the Imperial Valley, beholding the fruit from the seeds the settlers had sewn. "And

indeed it is a miracle, the transforming, within a dozen years, of a tract of strict desert into a farming region of the highest fertility."[5]

Transforming indeed. Imperial Valley was an imagined and engineered environment, a region refashioned for irrigated farming. The idealized portrayal showed hearty white pioneers sallying forth to take up the challenge of reshaping the desert wilderness into a commercialized Garden of Eden. Founders promoted a paradise of lush green rectangular fields worked by white hands and bearing fruits, vegetables, and grains to feed a hungry nation. Colonizers believed they were rejuvenating the nation's rural horticultural heritage and being faithful to Thomas Jefferson's famed yeoman ideal. Desert farmers conceptualized land and water as commodities to be altered and controlled for economic, racial, nationalistic, and psychological reasons. An unreliable "rampaging" Colorado River had to be "subdued" because it threatened crops, farmlands, and towns. Gullies, dirt mounds, and shifting sand dunes required leveling to get the "wildness" out of them. The Colorado River had to be "harnessed" by white Americans rather than Mexicans because North Americans would put it to "beneficial use" rather than "wasting" it as it flowed into the Gulf of California. As a senator from Arizona succinctly stated: "I want to see the third greatest waterway of the country comprehensively, systematically developed, its destructive forces subdued, there (*sic*) vast possibilities converted to the uses of man."[6]

In 1921, Winfield Barkley, a bank manager in the Imperial Valley, epitomized the language commonly used in describing the desire for environmental conquest. A poem titled "Wrestled with the River," published in a local paper, reminded readers of the difficulty of settling the desert, while celebrating its transformation. "Water, flowing down immense irrigation ditches for hundreds of miles is the key which has unlocked this reincarnation of the Garden of Eden." He urged his readers to "pause and marvel at the master minds which [en]visioned, amid desert sands, the empire of today." These pioneers "had the pluck, energy and engineering skill to make it a reality."[7]

Boosters of Colorado Desert settlement chose the term "Imperial" to reflect their racial, cultural, and nationalistic aspirations. As American landowners quickly took possession of the southern half of the valley in Mexico, the international border receded into the background and settlers and boosters began referring to the region as the "Imperial Valley of California" and the "Imperial Valley of Mexico," as if the basin was a whole unit bisected by an invisible line.

Imperialistic language served as therapeutic remedy to the suffering and toil expended in battling the harsh environment. This belief was expressed in mediums as diverse as eulogies, news stories, magazines and books, speeches, songs, movies, and poetry. The intent to subjugate the Colorado River and the natural world derived from a peculiar mental inclination. In time, curbing nature's destructive power laid the groundwork for the unrelenting quest for total control over nature, workers, and social relations. In sum, for the white settlers, racial and cultural supremacy legitimized the conquest of the landscape to bring forth an agricultural bounty. Building on how the settlers envisioned their quest to transform the desert into an irrigated paradise, this book explores interrelationships of power in the Imperial-Mexicali Valley from 1900 to 1940. In particular, this study examines the capitalistic transformation of nature for industrial agriculture, the creation of a racialized, hierarchical workforce and society, and the manner in which farming elites handled social and labor "agitators."[8]

Bisected by the international border, the Imperial-Mexicali Valley possesses the most inhospitable climate in the western hemisphere. It yields three inches of rain per year, with temperatures reaching eighty degrees in winter and 118 degrees in July, accompanied by blinding sandstorms. North of the border, the valley is forty-five miles long and thirty miles wide, with gradually rising slopes or mesas on the eastern and western sides. The Coastal Mountains flank its western edge and the Chocolate Mountains form its northeastern boundary. At one time, the basin lay under the Gulf of California. Immediately south of the Chocolate Mountains are sand dunes twenty miles wide, forty miles long, and up to sixty feet in height. These dunes are the remnants of windblown sand from the bottom of this ancient seabed. Arizona lies sixty miles to the east adjacent to the Colorado River.

The muddy river entered this locale after a 1,700-mile journey. For eons it carried silt and debris from seven states into the Colorado Desert. As the river flowed along the eastern rim of the valley, the sand dunes blocked its path into the basin and forced it southward to the gulf. Incidentally, this section of the international border sits at sea level, so once the river passed the sand dunes it

Lower Colorado River Basin, 1905. Courtesy Inventory of the Joseph Barlow Lippincott Papers, 1882–1942, Box 82, item no. 1057, From the Holdings of the Special Collections and Archives, Water Resources Collections, UCR Libraries, University of California, Riverside.

continued for miles before disgorging silt and debris across a vast delta plain. Over time, the elevation of the delta rose, forming a cone that blocked water from the gulf from entering the basin. Sediment and rocks piled along the bottom of the river channel, raising it higher every year until water ripped through its bank and cascaded into the valley. The fugitive water traversed westward through overflow ravines until it reached a depression in the basin floor 280 feet below sea level. During flood season, the river deposited topsoil, in some cases hundreds of feet deep—perfect for farming.[9]

Settlement began when a colonization company dug an aqueduct from the Colorado River in California, for fifty miles through Mexico, before turning back into the United States.[10] Settlers encountered very few indigenous residents. Ten extended Cucapá families lived in the Colorado River delta. Before white settlers arrived, during flood season, Indians from Arizona and the Coastal Mountains ventured to the desert to hunt, fish, and farm, and left when the ponds and creeks evaporated.[11] Drawn by widely disseminated publicity, homesteaders from around the nation snapped up almost half a million acres of the public domain. Control over land and water formed the basis of intensive market-driven irrigated agriculture.[12] Possession of these natural elements spurred growers to use the state to procure inexpensive, plentiful American and foreign migrant workers. Farm leaders rigidly controlled the racially hierarchical workforce. Believing they lived in a world of turmoil and chaos, settlers sought to eliminate social, political, and economic challengers by means of beating, jailing, and expelling malcontents and dissenters. Over the years, state mandates covering wages and field housing, as well as school attendance dictates, provoked distrust of outsiders and of government regulations.

Subordinated and marginalized individuals and groups contested the imposition of authority from the farming elite, while jousting among themselves for social and economic position. Resistance came in many forms, from violence to abandoning jobs at harvest time, union organizing and picketing, and criminal activity.[13] Over time, the highly stratified, racially diverse working class formed social and work networks. Labor coalitions attempted to bridge racial divides and reshape social relationships.

This book makes five contributions to the historical literature of early twentieth-century western irrigated agriculture, labor relations, and environmental change on the US-Mexico borderlands. First, it broadens our understanding of irrigated desert agribusiness. The valley's ideal weather permitted year-round farming of dozens of crops. Farmers on both sides of the border introduced large-scale cotton production on the West Coast, and the Imperial Valley led the way in boutique garden crops. Intensive farming, oftentimes growing two crops annually on the same ground, deposited millions of pounds of salt every year that had to be leeched from the soil. Moreover, to maximize profits, farmers sprayed millions of pounds of insecticide and pesticides, which fouled the land and the airshed.[14]

Poisoning the ecosystem infuriated Ella Ruth Johnson, a housewife residing in the northern end of the valley. In 1939, Johnson sent a letter to the state director of agriculture demanding to know why the county horticultural commissioner illegally permitted airplane crop dusters to spray calcium arsenate on pea fields adjoining her ranch. Breathing the poison sickened her and a

J. Edgar Ross tarred and feathered in Brawley for refusing to purchase a World War I Liberty Bond, October 1918. Courtesy s7-f3, Imperial County Historical Society, Pioneers Museum, Imperial, California.

female neighbor, forcing both to seek medical attention. Officials disclosed Johnson's letter as the first formal complaint against airplane crop dusting in California. The furious housewife skewered the farmer, claiming a small sum was saved from airplane dusting, while safer insecticides and other forms of application were available. Johnson viewed "the future with alarm since we fear that we shall consume arsenic repeatedly in various ways," namely, breathing contaminated air, drinking from ditches where poison had drifted, and eating fruits and vegetables saturated with chemicals.[15] The chemicalization of air, land, and water, retorted agribusiness spokesmen, was a regrettable outcome of industrial farming.

This project also expands our knowledge of how settlers attempted to subjugate and gain water rights to the Colorado River. Most noteworthy is the argument that water taken from the Colorado River for irrigation in the Imperial Valley has always been under federal control—first under the legal ownership of the Mexican government by virtue of a 1904 contract with the colonization agency that stated the canal was solely a transporting vehicle, for the agreement gave the colonization company no rights to the river, and later, when the US Department of the Interior finished construction of the All-American Canal in 1942 to move water through an aqueduct north of the border, the United States maintained ownership of the waterway and thus legal claim to the river.

Another important point is how farmers on both sides of the international line experimented with hiring a variety of foreign and domestic workers to establish a racialized, hierarchical agricultural labor force. Therefore, changes in the Imperial-Mexicali Valley labor market resulted in an international interracial borderland.[16] How groups individually and collectively resisted racism, discrimination, and second-class status is another noteworthy element of this project. Mexican migrants used the borderlands as a corridor to escape labor exploitation, while consistently demanding farmland in the Mexicali Valley. At the end of this story, campesinos compelled the government to expel the foreign landowners and redistribute the fertile farms to its citizens. Finally, this analysis delineates the cross-border relationships between the Imperial Valley and the Mexicali Valley. American landowners on both sides of the international border battled each other for control of the Colorado River, yet they cooperated to develop and market cotton, quarantine against crop pests, and recruit transnational Mexican workers.

Organized into this introduction, five chapters, a conclusion, and an afterword, this book begins with a discussion of how settlers and the US government "harnessed" the Colorado River to establish desert irrigated agriculture. As the water system was a transnational environmental enterprise, a series of Mexican governments wielded authority over water delivery and levees. The chapter summarizes congressional efforts to build the All-American Canal and Boulder (later renamed Hoover) Dam to prevent flooding and release water during dry spells, to remove excessive silt, and most importantly, to wrestle control of the waterway from Mexico. Significantly, local voters approved every major water-related decision, including granting federal ownership of the All-American Canal.

Chapter 2 focuses on colonization on both ends of the valley and the transformation of the environment for agriculture. Since landownership formed the foundation for legal and political power, this chapter discusses the demographic makeup of the population. It also explores the ideology and manifestation of desert conquest. Similarly, community leaders enforced racial segregation and mechanisms for social control. This chapter notes the myriad ways that people of color and working-class whites challenged social proscriptions. It also traces the development of scientific irrigation, crop development, and state and federal assistance to eradicate insects and rodents. The evolution of farming, absentee landownership, racial tenancy, and state assistance for crop development and "pest" control is the subject of chapter 3. It traces the integration of agribusiness associations, railroad shippers and distributors, and the role of the state in supporting industrial farming. Of importance is the orientation away from family farms and small-scale tenancy toward large-scale operations with the explosion of cotton acreage during World War I. After the postwar recession wiped out the cotton industry in the Imperial Valley, farmers switched to a mixture of profitable vegetables, dairies, and animal feed crops. Landowners began leasing to Asian tenants to grow labor-intensive fruits and vegetables. Meanwhile, across the border, the Colorado River Land Company (CRLC), owned by Los Angeles shareholders, purchased over 800,000 acres. After failing as a cattle ranch, the CRLC grew cotton using migrant Chinese tenants and pickers. By 1922, many onlookers proclaimed it was the largest cotton farm in the world.[17]

Chapter 4 traces the establishment of a segmented, racialized migrant labor market. Settlers initially relied on single white males and Indians to clear land, dig ditches, and set up towns. Transitioning to cotton production brought the issue of scarcity of field hands to the fore, prompting employers to experiment with a diverse international labor force. After World War I, the development of fresh produce, especially cantaloupes and lettuce, ushered in an agricultural revolution. Growers and shippers constructed a racially tiered labor force, consisting primarily of Mexican transnational migrants for fieldworkers and white and Japanese labor in the packing sheds. Interestingly, Mexicali landowners, with the assistance of the governor, maintained a localized Chinese emigrant importation racket, while farmers across the international line established a similar emigration policy for Mexicans. Workers on both sides of the border endured harsh working and living conditions. During the 1920s, the California legislature entered into farmer-worker relations to enforce sanitary housing and labor laws and compulsory school attendance. A decade later, white and black American migrants, along with Filipino mobile laborers, saturated the labor market, driving down wages and forcing the state and federal governments to provide public assistance.

Impacts of field and packing-shed rebellions along the borderland are the subject of chapter 5. During the first two decades of colonization, workers frequently engaged in wildcat field strikes to increase wages. The entry of white radical union members of the Industrial Workers of the World in the 1910s launched the first major organized field labor challenge to agribusiness. Mexican workers organized massive strikes in 1922 and 1928 to improve working conditions and raise wages. Concurrently, white shed workers frequently went on strike to increase wages. During the 1930s, the interracial agricultural labor force cooperated in enormous field walkouts. Communist union organizers and members of the American Civil Liberties Union aided in unionization and strikes. Corporate agriculture in California responded in kind by organizing the Associated Farmers, which teamed up with US labor conciliators and the Mexican consul to bring workers to heel. After enduring withering repression, disgruntled Mexicans crossed the border and demanded farmland from the government, setting the stage for a demographic transformation of the Mexicali Valley. A concise conclusion is followed by an afterword to update the themes in this study after 1940 and suggest recommendations for addressing the region's profound socioeconomic-environmental challenges.

CHAPTER 1

"SUBJUGATING" THE COLORADO RIVER

Water is King—here is its Kingdom

Imperial Press, 1901

When California senator Hiram Johnson urged his colleagues to pass legislation to build an "All-American" Canal and the world's largest dam on the Colorado River, the famed wordsmith framed the issue as a struggle to conquer nature. "We propose to make this river, storm and devilish in character, a servant to mankind and the territory through which it flows." Three years later, in 1930, at the groundbreaking ceremonies for Boulder Dam (later renamed Hoover Dam), the secretary of the interior considered the slab of concrete soon to hold back the mighty river as "one of man's greatest victories over nature." To government engineers, the dam symbolized the subjugation of nature for profit. "These controlled waters," the secretary of the interior promised, "will care for millions of people and create billions of wealth." When the dam was completed in 1935, the *Brawley News* trumpeted: "The great Colorado river has gone on its last rampage. No longer will it threaten Imperial valley with floods and never again will it break over its levees, deluge the ranch lands with surplus water, or cut any more new rivers in a wild effort to reach the Salton sea." Impounding the water behind the concrete wall had other benefits. "Neither will there be any more droughts such as was experienced here last year. The great stream is now under absolute control—all due to the conquering energy of man."[1]

This chapter lays the foundation of power and control in the Imperial Valley by examining the audacious effort to divert and manage the Colorado River for agriculture. It begins with dreamers and boosters who envisioned an irrigated desert Garden of Eden. Settling the northern half of the valley, some half-million acres of the public domain, depended on diverting the river in a canal for fifty miles through Baja California. The enterprise involved a dizzying cast of actors: the California Development Company (the colonization agency), a Mexican landowner of the Mexicali Valley, the Mexican government, American settlers, an interracial workforce, the Southern Pacific Railroad, and the US Reclamation Service. The middle section examines a series of floods that almost ended colonization. These floods created the Salton Sea,

California's largest inland body of water. In the midst of building levees in the Mexicali Valley to hold back the river, Mexican revolutionaries invaded the northern district of Baja California, threatening to destroy the water system. The political crisis militarized the borderlands as US and Mexican troops hunted down filibusters and revolutionaries.

In the wake of the floods and amid a thicket of lawsuits and political instability in Mexico, settlers rejected private ownership of the water system and created the Imperial Irrigation District (IID), a public utility. The success of public ownership of the water infrastructure on both sides of the border led to an alliance between the IID and the Reclamation Service to build the All-American Canal, an aqueduct wholly in the United States. Settlers began a decade-long campaign to convince the US Congress to "subjugate" the river with a government-built dam and aqueduct. Wealthy southern Californians who owned most of the Mexicali Valley opposed the new aqueduct for they rightfully feared Imperial Valley farmers would monopolize the river. Thus landowners on both sides of the international line rushed to put as much ground into production as possible to inflate a historic water right to the Colorado River.

California Development Company

In the first half of the nineteenth century Americans sailed up the Colorado River from the Gulf of California or crisscrossed the Colorado Desert on their way to the coast or to map the international boundary. Travelers wrote of the lack of water, the heat, and the barrenness, leading readers to deem the region uninhabitable. The trek through the desert, known as the Emigrant Trail, was easy to follow because of the broken wagons, bones, and trash. A few enthusiastic souls lobbied Congress for permission to colonize the region, but their efforts came to naught. In the last quarter of the nineteenth century, western migration, a growing demand for food, and the growth of large-scale agriculture in California dovetailed with lenient land grant policies and a wilderness reclamation movement to crystallize the idea of reclaiming the Colorado Desert for irrigated farming. Once the railroad skirted the northeastern section of the basin in 1879, capitalists envisioned greening the desert.[2]

In typical boom-and-bust western fashion, boosters pitched proposals to draw water from the Colorado River to colonize various portions of the Colorado Desert. So began a tangled web of intrigue, betrayal, euphoria, and despair. Our story begins with Charles Rockwood, who dropped out of engineering school at the University of Michigan due to failing eyesight. Rockwood headed west, bouncing around for a decade taking survey and irrigation engineering jobs. In 1892 his life changed. Directors of a speculative venture hired him to survey a route to divert the Colorado River east to irrigate a million acres the firm considered purchasing in Sonora.[3] Rockwood found the project unrealistic and turned his attention instead to irrigating the desert valley in the United States. He planned to go around the enormous sand hills blocking his path by digging a looping canal around the sand hills south of the border and then back into the United States. In 1893, his partners agreed to the plan and set out in search of funding. Rockwood and the company's president traveled both sides of the Atlantic pitching the plan to investors. The company soon ran out of money when a recession hit and Rockwood traded the survey maps and equipment in return for back wages for himself and his survey crew. Rockwood found new investors, including William Heffernan, a surgeon on the Yuma, Arizona, Indian reservation, near the area where Rockwood planned to divert the water from the river. Heffernan purchased the company at a sheriff's auction.[4]

In order to secure the land through which the canal passed in Baja California, the company paid for an option on a quarter of a million acres owned by Guillermo Andrade, the Mexican consul in San Francisco. Rockwood and his partners agreed to pay an installment of $5,000 and $250 a month until the land could be purchased. After the first option expired, Andrade renegotiated an option on 100,000 acres for $1,500 and 60 cents an acre for the fifty-two-mile canal right of way through his property.[5] Next, the company put a down payment on the property in the United States where the prospective canal would cross the short distance before it entered Mexico. Now in nominal control of the land underneath the proposed canal, the company followed California law requiring posting a notice to divert a specific amount of water from a stream and record the action with the county recorder. Every ninety days in 1895 a company director nailed the notice on a tree next to the river. In addition to state law, federal law required permission from the War Department to divert water from the Colorado River because it was a navigable stream under a treaty with Mexico. Angling to avoid scrutiny from the US and Mexican governments, the company proceeded without informing either government. For a fresh start, the partners chartered the company in New Jersey and changed its name to the California Development Company (CDC) in April 1896.[6]

Rockwood also had to get around a Mexican law prohibiting foreigners from owning land and businesses along its borders. When a Mexican court threatened to nullify the deal made with Guillermo Andrade because the CDC was a foreign company operating in Mexico, the CDC created a Mexican subsidiary named La Sociedad de Irrigación y Terrenos de la Baja California, S.A. (La Sociedad). Lacking funds, the company renegotiated its agreement with Andrade, who received company stock as partial payment on 100,000 acres. The deal also required that half of the water in the canal be made available to farmers south of the border. Despite the progress, the company struggled to attract investors. In addition to liens, office and travel expenses, and options on Andrade's property, New Jersey threatened to cancel the corporate charter for failure to pay a renewal fee and taxes.[7]

When all hope seemed lost, the CDC had two strokes of luck. William Smythe published *The Conquest of Arid America* (1899), popularizing irrigation farming in the arid West. He wanted the government to "harness" rivers for irrigation to homestead desert "wastes." Smythe's messianic prose promoted a new Eden by subjugating rivers and civilizing the West's desert "wilderness." He dedicated the book to "homeseekers who, under the leadership of the paternal Nation, are to grapple with the desert, translate its grey barreness [*sic*] into green fields and gardens, banish its silence with the laughter of children." The conquest of arid America, claimed Smythe, was a nationalistic task. He challenged white Americans to fulfill their patriotic duty to populate the nation's deserts. In return, hearty pioneers would realize personal satisfaction and strengthen the nation.[8]

A master marketer, Smythe strove to modernize Thomas Jefferson's vision of the western lands populated by white landowning farmers, the backbone of democracy and capitalism. Private colonization of the arid West, he charged, had created massive estates and warped Jefferson's vision of yeoman family farmers. "The essence of the industrial life which springs from irrigation is its democracy. The first great law which irrigation lays down is this: There shall be no monopoly of land." He brokered no patience for land barons for they threatened the fabric of American life. "A large farm under irrigation is a misfortune; a great farm, a calamity. Only the small farm pays. But this small farm blesses its proprietor with industrial independence and crowns him with social equality. That is democracy." *The Conquest of Arid America* made Smythe a national figure. His message merged with lobbying from other irrigation advocates, resulting in the passage of the Reclamation Act of 1902 to promote government-planned irrigation projects in the arid West. Over the next decade Smythe played a major role in supporting the federal takeover of the CDC's water system.[9]

George Chaffey

The CDC finally hit pay dirt in late 1899 when it persuaded George Chaffey, an engineer with successful irrigation projects in California and Australia on his resume, to bankroll the Colorado Development Company. The company's directors concealed the company's financial predicament as Chaffey examined the terrain and engineering plans. After initially declining to come on board, he later reconsidered. Led by an elderly Cucapá guide, Chaffey spent two months mapping a new diversion course before agreeing to build the water system. Since this was an irrigation project, Chaffey separated his responsibili-

Steamboat on the Colorado River near the international border, ca. 1906. Courtesy Inventory of the Joseph Barlow Lippincott Papers, 1882–1942, Box 82 item 2052, From the Holdings of the Special Collections and Archives, Water Resources Collections, UCR Libraries, University of California, Riverside.

ties of managing the water delivery system from the risky business of land promotion and colonization. In March 1900, the CDC established the Imperial Land Company to advertise and facilitate public land filings, carve out town sites, and develop urban water and power services.[10]

The fifty-two-year-old Chaffey had a proven record and showed so much self-confidence he did not examine the company's books before signing on. In fact, the books were inaccessible because the CDC could not pay its lawyers and accountant, who refused to give them up! When Chaffey was informed of the company's financial status, he demanded a new contract. In return for his personal investment of $150,000 Chaffey became the company's president, general manager, and chief engineer. The agreement gave Chaffey control of the CDC for five years through a majority of proxy votes of company stock. Once the canal was built and the project became profitable, he was entitled to stock, cash, and water rights.[11] It was a humiliating turn of events for the CDC stalwarts who had worked tirelessly for years and had exhausted their personal funds. Now they had to give up management of the company to the interloper who doubted their abilities and integrity. Chaffey created a new company, Delta Investment, to distance himself from CDC debt. Chaffey put the project on sound footing by transferring CDC bonds to the new company, paying off creditors, and buying out earlier investors. Then he funneled his money and that of new investors into Delta Investment.[12]

With the financial house in order, Chaffey devised a master plan to water the Colorado Desert. He understood the herculean task of supplying water to a half million acres would be costly, time consuming, and an organizational nightmare. So he could focus on the canal system in Mexico, Chaffey established private water companies (mutual water districts) to distribute water to farms. Chaffey charged the water companies 50 cents for every acre-foot (enough water to cover an acre with one foot of water) it delivered. The water companies billed farmers a delivery fee to pay for the lateral canals, maintenance, and overhead; in addition, farmers had to buy one share of water stock for every acre of land they owned. Thus an acre of land was worth what the water companies could get for stock, about $12 a share.[13] Chaffey oversaw this colonization scheme through Delta Investments, which controlled CDC, La Sociedad, Imperial Land Company and, indirectly, the mutual water companies. On Thanksgiving 1900, workers opened the river headgate and water flowed south a hundred yards before crossing into Baja California. The canal ran parallel to the river for four miles before turning right, and the muddy stream slogged westward through a ravine Chaffey renamed the Alamo Canal. As dredgers straightened and widened the canal, workers built a headgate south of the international border. On a blistering hot day in June 1901, Chaffey opened the headgate and water flowed into California.[14]

Chaffey needed to establish a port of entry into and out of Mexico for the payment of Mexican and US customs duties, so he filed on a 320-acre desert homestead near where his canal entered the United States and donated it to Imperial Land Company to set up a town site. Envisioning the joining of two nations in his binational environmental enterprise, Chaffey intermixed the names of California and Mexico to name the twin towns of Calexico on the American side and Mexicali on the Mexican side of the invisible international line. The energetic engineer and clever marketer had other creations in mind. He discarded the bleak name "Colorado Desert" and replaced it with the "Imperial Valley," signifying his grandiose dreams. But the new appellations concealed more than they revealed. In songs, poems, autobiographies, and histories of colonization, sturdy white settlers conquered the river and the desert. In reality, indigenous residents were the principal canal and farm labor force for the first decade.[15]

When Americans began cattle ranching and plying the lower Colorado River delta in steamboats fifty years earlier, the Cucapá had resided there for a millennium. Consisting of ten extended families totaling hundreds of inhabitants, they resided in rancherías consisting of a dozen homes. Cucapá lived a subsistence lifestyle, fishing, hunting, and farming in the fertile delta floodplains. Trading networks brought material goods and food. They served as guides and laborers on the steamboats and cut wood for fuel. The introduction of the wage economy profoundly impacted the tribe. Over time, they incorporated American-style trade goods, food, and clothing. When the Southern Pacific Railroad arrived in Yuma, Arizona, in 1877, the steamboat era ended. Unemployed Cucapá lost jobs, forcing many to leave the region. Some moved to Yuma, while others took jobs with the CDC digging the Alamo Canal. Later, they cleared land and built homes.[16]

While Chaffey moved the project along, Charles Rockwood, the former chief engineer, plotted to remove him. Rockwood bitterly resented Chaffey's success. Chaffey had also punished Rockwood for failing to inform him of the CDC's shaky finances by demoting him to an assistant engineer of a water company. Although technically Chaffey controlled the CDC through a majority of CDC shares, he neglected to put them in a trust. Rockwood exploited this error by convincing shareholders to allow him to carry their proxy votes. In this way, Rockwood ousted Chaffey in February 1902. Rockwood's allies leveraged Chaffey's success by getting a loan to buy him out. Chaffey left with the money and company stock (eventually worth $100,000). Rockwood lured the petulant Anthony Heber, the CDC's original president, back to manage the company, while Rockwood assumed the post of chief engineer.[17]

Rockwood and his associates had little time to celebrate. Keenly interested in the irrigation farming prospects in the Imperial Valley, the Department of Agriculture sent investigators to conduct a soil investigation. The soil report declared the region saturated with alkali and only portions fit for farming. The widely published findings tarnished the CDC's reputation and ruined its credit. The federal government had other reasons to discredit the colonization scheme. A CDC surveyor drawing township and range lines discovered the government's US-Mexico boundary surveys were off by one to one-and-a-half miles, and the border monuments had been haphazardly placed, moved, or were nonexistent, throwing into question thousands of acres of homesteads. At the same time, federal officials received reports of violations of the homestead laws. While Congress passed a bill for a new border survey, the secretary of the interior halted land grant filings and ordered an investigation into homestead frauds.[18] The Interior Department bought time while the Reclamation Service devised a plan to take over the irrigation project.

Reclamation Act of 1902

Transporting water to the Imperial Valley was part of a larger effort by nationalists and visionaries at the turn of the century to control and "harness" the power of rivers to irrigate the arid West. Proponents of western settlement framed the subjugation of nature in moralistic and nationalistic terms.[19] Landscapes like deserts unfit for human use needed to be redeemed from nature. Referring to a flood in 1891, the secretary of the Historical Society of Southern California wrote: "The Colorado River seems to have repented of its evil work, and is now seeking to atone for its great sin, in desolating so large a portion of the earth, by refilling the desert sea."[20] Opponents of private colonization campaigned to persuade the federal government to build reclamation projects in the desert. Congress responded to the desire for government-assisted irrigation with the 1902 Reclamation Act. The newly created Reclamation Service (later renamed the Bureau of Reclamation), within the Department of the Interior, would use monies from the sale of public lands for irrigation projects. The act required landowners to live near or on land that received water from a federal project. Furthermore, to prevent land monopoly, it forbade ownership of over 160 acres on federally financed reclamation projects.[21]

California Development Company officers greeted the passage of the act with alarm, for they feared a rival from the Department of the Interior. Moreover, the Department of War (which held authority over the international river) could block river diversions without federal approval. The first challenge, however, came from Mexico. Five months after water crossed into the Imperial Valley, the Mexican government sent a diplomatic note protesting the diversion in violation of several treaties protecting the navigability of the river for commerce. Alarmed that Mexico could potentially halt development of the Southwest by blocking substantial water diversions, the United States sent a researcher to examine the situation. Predictably, the investigation yielded a report determining the canal would not impact the river's navigability. Although dependency on water placed colonization in the hands of the Mexican government, that dependency also offered the CDC a shield of protection against its rival, the Reclamation Service. When the Reclamation Service told the irrigation company it did not have federal permission to divert the river, its officers defiantly replied it had an agreement with a Mexican landowner to take water from the stream, and the Alamo Canal fell under the jurisdiction of La Sociedad.[22]

An incredulous secretary of the interior put the breaks on this colonization project by delaying the boundary survey and investigating allegations of Desert Land Act fraud. Federal investigators discovered cases where Imperial Land Company hired phony entry men to file on homesteads and then sell the land to Imperial Land Company, who turned around and sold the property. The investigation resulted in the closure of the Land Commission office in the Imperial Valley. To punish the settlers, the government refused to finalize homestead property titles for five years. In a ploy to seize control of the project, the Reclamation Service used the previously mentioned soil report to discredit colonization. Settlers proved the government wrong when they showed miles of crops to visiting newspaper editors.[23]

1905–07 Floods

An unpredictable Colorado River overshadowed the CDC's political tribulations. Providing water to farmers proved more difficult than anyone imagined. Silt and sand blocked the canal intake, forcing overworked dredgers to clear the path. Even more alarming, low river levels in the winter of 1903 and 1904 caused an irrigation drought in the valley. When farmers threatened to sue the CDC for nondelivery of water, engineer Charles Rockwood cut bypasses around the jammed intake, but these also filled up. In desperation, Rockwood picked a spot south of the border and ordered a dredger to rip a hole in the riverbank to get water into the Alamo Canal. The CDC's decision to divert water from the riverbank without permission from the US and Mexican governments would later have profound consequences. Rockwood's frantic effort failed and farmers rained a flurry of lawsuits on the water company.[24]

The Reclamation Service pounced at the opportunity to wipe out the California Development Company. Earlier, technocrats announced a plan to build four dams, electric generators, and canals along the river. Seeking to undermine the CDC, the Reclamation Service posted a water diversion notice giving it control of the Colorado River. The Reclamation Service's power play anticipated luring Imperial Valley settlers to join the government's water project in Yuma, Arizona, on the other side of the river. In February 1904, CDC president Anthony Heber responded to the government's challenge by announcing the CDC's doubling of its original claim from the river. When the Departments of Interior and Justice pushed the argument that the river's navigability was in danger, the CDC persuaded the California delegation to submit a bill stating that irrigation superseded navigation.[25] Anthony Heber boarded a train to Washington, DC, to support the bill. At the hearing, William Smythe, the famous irrigation booster, blasted the CDC for violating the letter and spirit of numerous land and water laws. Heber did his best to defend the CDC, but it was a losing proposition. Under heated questioning, Heber grew defiant: "It is my earnest desire to worship at our own altar and to receive the blessing from the shrine of our own Government, but if such permission is not given we shall be compelled to worship elsewhere." After scuttling the bill, the Reclamation Service and Smythe rushed to link the Imperial and Yuma projects.[26]

Under William Smythe's guidance, landowners formed the Imperial Valley Water Users Association and held rallies in support of government reclamation. The plan called for the Water Users to incorporate for $12 million and buy the privately held mutual water companies. Then the Reclamation Service would purchase the CDC. Government engineers estimated an aqueduct across the sand to cost $20 million.[27] To beat back the Reclamation Service's challenge, Heber and associates traveled to Mexico City, and with their well-connected lawyer hammered out a deal with the Mexican government recognizing the CDC's Mexican subsidiary La Sociedad de Irrigación y Terrenos de la Baja California as the water conveyance entity managing the Alamo Canal. It was a fabulous deal for the CDC and for Mexico. This is an important point, since Mexico, at any time, could have revoked La Sociedad's permit to transport water. In return for the concession, La Sociedad agreed to sell up to half of the water in the Alamo Canal to Mexicali Valley customers. Mexico's Congress enshrined the deal into law. From this point forward, Mexico exerted authority over La Sociedad's water distribution system.[28]

The CDC-Mexico agreement torpedoed the Reclamation Service's takeover of the irrigation project. Yet it came at a steep price, as Heber agreed to sell water to farmers for the development of the Mexicali Valley. This decision had profound ramifications as agriculture in the Mexicali Valley expanded in the 1910s and Imperial Valley farmers felt threatened because the Mexicali Valley was twice as large as the Imperial Valley, which potentially meant a dramatic decrease in water for US farmers. It is critical to emphasize that the CDC viewed itself as a transnational water company, while Imperial Valley settlers were nationalists and preferred not to share the river with Mexico. Shrewdly, Heber retained Alberto Andrade to manage La Sociedad. Alberto was the son of Guillermo Andrade, who formerly owned the Mexicali Valley and had sold the CDC the right-of-way for the Alamo Canal. With Alberto Andrade as a partner, the CDC kept favorable relations with the Mexican government.[29]

All of the deal making, however, did not produce enough water from the Colorado River to satisfy farmers. This forced engineer Rockwood to cut another unprotected gap in the riverbank, but the intake quickly silted up. Stalling for time while dredgers cleared the muddy intake, Heber put the company up for sale for $5 million in July 1904. Settlers countered with $1.25 million. After agreeing to arbitration, Heber lowered the offer to $3 million. In August, the Water Users Association accepted Heber's asking price. William Holt, the most prominent local businessman, went to Washington, DC, to convince the Reclamation Service and President Theodore Roosevelt to support the buyout.[30] The California Development Company's greatest adversary was not the Water Users, the Reclamation Service, the departments of Interior or War or even debt. The river proved the colonizer's primary nemesis. Panic stricken because of low water levels, Heber and Rockwood decided to open another gateless intake (without Mexico's permission). The CDC's superconfident leaders ignored numerous reports regarding the river's indomitable force at flood stage, believing they could quickly build a headgate. In September, Rockwood dredged a fifty-foot-wide gap in the riverbank.[31]

When news reached Imperial Valley of the government's willingness to accept the $3 million offer, Heber called a meeting in Brawley, where he announced to the stunned audience that the CDC was no longer for sale. When Paul Van Dimas, a leader of the Water Users Association, tried to make his case, Heber blasted him and the idea of government ownership. On cue, a crowd seized Van Dimas and gave him a bath of hot tar and feathers before tossing him on a train bound for Mexicali. Back in Brawley, the CDC held a barbecue and entertained the crowd with fireworks. The turn of events infuriated President Roosevelt, and he ordered yet another investigation of colonization activities.[32] In light of the uncertainty surrounding the whole scheme, the secretary of the interior advised against purchasing the canal company. Instead, he categorized the Colorado River a navigable stream, rendering the CDC's water filings under California law null and void. In typical bureaucratic contradictory fashion, the secretary also recognized the farmers had put the water to beneficial use, so he would not block water from being diverted from the river. In January 1905, the first of a series of flash floods unclogged the two jammed intakes, sending a torrent of water into the valley.[33] Desperate for cash to build a headgate, Anthony Heber asked the Southern Pacific Railroad for a $200,000 loan. The railroad turned him away. Weeks later, behind Heber's back, Charles Rockwood and other directors approached the railroad for $200,000 in return for majority control of the CDC, as well as management authority.[34] Recall, just a few months earlier the government offered to buy the CDC for $3 million.

While railroad mogul Edward Harriman considered the proposal, a third flood came in March 1905. Rockwood ordered the hole in the riverbank lined with wooden pilings and plastered with flimsy brush like a Band-Aid. Workers piled thousands of sandbags to hold back the river. Two more floods obliterated the protective work and raced down the Alamo Canal across the border to the Salton Sink and flooded the New Liverpool Salt Company.[35] The ignoble colonization scheme was over. In early 1905, Charles Rockwood and allies secretly agreed to sign over the CDC to Harriman at the company's annual meeting in New Jersey in June. This was the second time Rockwood had double-crossed a CDC president. The magnitude of the transaction was breathtaking. For a $200,000 loan, Harriman got a majority of CDC stock (controlled by a trustee as collateral for the loan—a move that George Chaffey failed to make), a majority of seats on the board, and the right to appoint its president and general manager. Harriman now had management control of the Alamo Canal, the right to sell water to farmers on both sides of the line, and 100,000 acres and other assets held by La Sociedad.[36]

The railroad took control in June and had a major task before it. The hole in the riverbank had widened to 160 feet. The break in the riverbank attracted national attention, and no less than four-dozen engineers over the next two years opined on how to close the gap. The railroad instructed engineer Charles Rockwood to design plans to close the crevasse, which called for a jetty across the river to force the river away from the intake. The mighty river pulverized the jetty. Plan two called for a proper headgate to control the flow. Called away on business in August, Rockwood left the effort to a railroad engineer who decided, instead, to build another jetty. The river washed it away, and hundreds of laborers, horses, and wagons scrambled to escape the cresting water. The largest flood of the season hit in November, widening the gap to six hundred feet. When the river subsided it left a mud wall blocking the stream's path to the gulf. The river backed up and changed course, cascading through the unprotected hole in the channel.[37]

Water spread out a mile wide as it entered the valley, ripping gashes in the sandy soil and creating ravines and rapids. Panic-stricken settlers fled in terror. The rising Salton Sea forced the Southern Pacific Railroad to move its railroad tracks to higher elevation. By March 1906 most of the region was under water. Explorer and writer George Wharton James, accompanied by five white men and two Cucapá, floated down the river in a canoe from Yuma to the Salton Sea, as bug-eyed onlookers swore they were fools. It was a harrowing experience. Massive logjams, huge waves, and rapids buffeted the adventurers. Reporters, sightseers, and engineers rushed to behold the four-hundred-square-mile desert inland sea, a body of water as large as Utah's Great Salt Lake. The alarming reports of the inundation reached the Mexican government, who appointed a commission of engineers, led by General Ángel García Peña, to investigate. A few days after a monstrous flood on November 30, 1905, the commission of Mexican engineers took a steamboat from Yuma, Arizona, to the Alamo Canal. Curious to see the flood, engineer Edwin Duryea made periodic trips to the region in 1905. In a popular magazine, Duryea joined a host of critics who skewered the CDC for not building a headgate.[38]

Burdened by criticism for his engineering ineptitude, Charles Rockwood resigned. Railroad mogul Edward Harriman sent his best engineers to close the chasm and authorized spending another $500,000. Harriman picked Harry Cory, a former engineering professor and assistant manager of the Southern Pacific, to close the gap. Cory looked out and saw the river spread across the desert seven miles wide. Waterfalls over forty feet high carved two canyons forty miles long on opposite sides of the valley. Although a six-foot sandbag wall saved Calexico, the waterlogged soil beneath Mexicali gave way and it slid into the stream. Settlers beseeched federal officials for help. Harry Cory estimated the river washed away the equivalent of four times the amount of earth dredged for the Panama Canal.[39]

With the full backing of Harriman's railroad empire, engineer Cory went to work in July 1906. After ordering a railroad spur built to the worksite, he commandeered three hundred side-dump cars known as battleships to transport rock. Each had a capacity of fifty to sixty tons. He also sent for a dozen work trains. The CDC's ragged tugboats and steamers lent support. Cory had

Mexican Colorado River Commission on steamboat in Alamo Canal, ca. 1906. Courtesy Inventory of the James D. Schuyler Papers, bulk 1886–1912, Box 2, folder 11.8, item no. 1, From the Holdings of the Special Collections and Archives, Water Resources Collections, UCR Libraries, University of California, Riverside.

access to quarries within four hundred miles. He also brought in a small forest of timber, another of cable, hills of coal, and mountains of food and supplies. Nature had met its match. All Cory needed was strong backs and men willing to work under extreme heat and dangerous conditions. At first, the railroad used white workers, but they quickly abandoned their jobs. When Cory's plan to import five hundred Mexicans fell through, the railroad contacted the US Indian outing agent for assistance. The agent ordered four hundred Native American men and their families from six nearby tribes on both sides of the border to the site. Native residents probably had little choice in the matter, but they were paid, and the flood threatened their homes and lives as well. The railroad also hired six hundred white men to work in a nearby quarry. Native Americans earned 20 cents an hour, while white laborers took home 27 1/2 cents to 30 cents an hour. Alberto Andrade, the head of the California Development Company's La Sociedad and the assistant general manager of the Southern Pacific Railroad in the Mexicali Valley, ferried material and workers to the worksite.[40]

Underpaid and segregated native workers saved the project. Years later, Otis Tout, a local newspaper editor and the author of a settler history of the Imperial

Valley, gave the indigenous people their due: "After searching for labor supply in many places the only source seemed to be the Indian tribes of Arizona and Mexico; through Government agencies practically all the men, women and children of six Indian tribes," two thousand in all, arrived at the worksite. "Pimas, Pagagoes, Maricopas and Yumas from Arizona, and the Cocopahs and Dieguenos from Mexico." Mexico declared martial law on the river and the tribal members lived in a segregated camp under the watch of Mexico's famed rural police. "Without this Indian labor supply," intoned Tout, "it is difficult to see what the engineers would have done." "All other labor was of an extremely transient nature. Hundreds of men came in on the supply trains and left again in a few days."[41]

For three weeks in August, the summer's hottest and most humid month, workers toiled around the clock to build a brush and wire mattress as a foundation for a rock dam. As the river receded, railcars rumbled across a rickety bridge dumping rock. For one month, day and night, rock crashed into the wire mattress. In October, the river obliterated the headgate, costing the railroad $122,000 and four months of labor. Undaunted, Cory redoubled his efforts. Six railcars every hour for three weeks dropped a total of three thousand loads of rock, gravel, and clay into the chasm. On November 4 the gap was sealed. A month later the river grew five times its normal size. The swollen river punched through the riverbank and washed away earthen levees. Some of the best engineers in the world had failed. They assessed the damage, considered the $1.5 million spent, contemplated what lay ahead to plug the gap ($350,000) as well as to build brand new levees ($1.2 million), and suggested that Edward Harriman pack his tent and walk away. He agreed.[42]

In mid-December 1906, Harriman telegrammed his decision to President Theodore Roosevelt. The press had widely reported the efforts to close the break; when the news reached Washington, DC, politicians expressed alarm. Crestfallen at the prospect of Harriman's desertion, settlers pledged to raise $950,000 to continue the effort. Concerned about the loss of life and property, President Roosevelt ordered Harriman to dam the breach, arguing the railroad was responsible for the disaster. After two weeks of negotiations, Harriman relented once the president promised to ask Congress to reimburse the railroad. In a message to Congress in January, the president lambasted the CDC and the Southern Pacific for gross incompetence. Nevertheless, he asked Congress for $2 million to reimburse the railroad for expenses incurred and to buy the CDC. Congress demurred on the grounds that public expenditures should not bail out private colonization.[43] The railroad committed all its resources to overpower the "rebellious" river. For a month, indigenous workers and Mexicans built trestles over the gap in the riverbank. After three trestles washed away, workers pounded two trestles into place and the entire southwestern freight system shuttled cars to the site. For two weeks, three thousand flat railroad cars carried 5,765 loads of rock, gravel, and clay from quarries five hundred miles away for workers to dump into the chasm before sealing it in February 1907.[44]

Next, Cory ordered the construction of over twenty miles of earthen levees to shore up the riverbank. A trainload of rock was parked nearby in case the river broke through. Workers along the river measured the flow and watched for flash floods. The railroad's telephone wires hummed as Cory's sentinels monitored the river. From start to finish, it took a year and a half to complete the task. It came in the nick of time because the summer flood of 1907 was the highest on record. The river constantly broke through levees, forcing the railroad to build miles of rock-topped dikes that had to be raised a foot each year to keep up with the rising riverbed. Although settlers, railroad executives, and reporters celebrated the "conquest" of the Colorado River, anyone who wanted to see the power of the waterway could feast their eyes on the Salton Sea. The colonizers' ineptitude also displaced the Cucapá. Flooding forced the indigenous inhabitants to move to Arizona and Sonora. The construction of levees years later allowed some Cucapá to return to the delta, where they took seasonal farm jobs. After damming the gap in the river, the railroad handed Congress a bill for $3 million. The cost was so high because settlers reneged on the $950,000 pledge for reimbursement. Congress refused to reimburse the railroad.[45]

Before the Reclamation Service conducted comprehensive engineering surveys of the Colorado River, the Mexican government studied the Colorado delta and gathered information about landowners and property boundaries. Like the US government, the Porfirio Díaz administration had little experience with large-scale irrigation projects, so Mexican engineers visited irrigation sites and attended conferences in the United States. In mid-1907, six months after the railroad closed the break in the river channel, a Mexican commission went to the National Irrigation Congress in Sacramento, California. Eager to curry favor with the commissioners, the Colorado River Land Company, the owner of most of the Mexicali Valley, hosted the commissioners in Los Angeles and then showed them around the waterworks in the Imperial-Mexicali Valley. The head of the commission, Fernando Beltrán y Puga, submitted a report to the secretary of foreign relations stressing the urgency for a treaty sharing the Colorado River and supporting storage dams in the United States.[46] The following year, Fernando Beltrán y Puga and a team of surveyors planned to visit the Mexicali Valley, and Beltrán y Puga asked the CRLC to procure supplies and laborers. Revealing the CRLC's political connections, in February 1909, Harrison Gray Otis, the CRLC's leader, met with President Porfirio Díaz to discuss company matters. Four months later, the CRLC complied with a request by the Mexican consul in Calexico for a map of property owners in the Mexicali Valley. The consul followed up with a request for a map of the Reclamation Service project near Yuma.[47]

During this period, the US government was antagonistic toward the Imperial Valley settlers and the railroad. Since the Interior Department refused to confirm land patents until 1909, settlers could not sell homesteads and found it difficult to secure credit. Moreover, the government's refusal to grant homestead patents meant the Southern Pacific Railroad could not collect on collateral notes or foreclosed mortgages. Frustrated with the financial losses, the railroad sued former CDC officers to return swindled stock. Rockwood and his associates returned $450,000 in stock. In 1909, when the river broke through levees and flooded the Mexicali Valley, the railroad refused to fix the protection works. Determined to protect their investment, CDC bondholders tendered a bid to buy the company. When the railroad refused to sell, in December, bondholders asked a judge to declare the company insolvent, and the court appointed W. Holabird, a high-ranking Southern Pacific Railroad official and personal friend of Edward Harriman, as the CDC's receiver. A Mexican judge appointed Alberto Andrade, the manager of the railroad in the Mexicali Valley and the general manager of La Sociedad, to continue as La Sociedad's receiver. Holabird sold bonds to build and repair levees.[48]

Although both the CDC and La Sociedad were under receivership, the railroad held the majority of its stock, and behind the scenes it played a complex shell game. After its lawsuit in Mexico against La Sociedad resulted in a $900,000 judgment in gold, in August 1910 the railroad moved to break free of the $2 million in legal claims against La Sociedad by forming La Compañía de Terrenos y Aguas de la Baja California, S.A. (La Compañía). After negotiating a fifty-year concession with the Mexican government to operate the Alamo Canal, in January 1911, La Compañía paid $325,000 to buy La Sociedad's assets and 70,000 acres. This transaction left La Sociedad essentially a company existing on paper with a mountain of debt and liabilities. Alberto Andrade became La Compañía's general manager, while continuing on as La Sociedad's receiver.[49]

In early 1910, when floods threatened to burst through dikes, receiver W. Holabird was powerless to act. Settlers and the railroad sent telegrams to Washington, DC, pleading for help. In June, Congress appropriated $1 million for levees in Mexico. The following month, President William Howard Taft selected John Ockerson to oversee the project. During the summer, Ockerson surveyed the situation, and that fall he met with federal officials to explain the project. To uphold its sovereignty, the Porfirio Díaz administration refused to permit Ockerson or any other American official to work in Mexico. After delicate negotiations, in December Mexico permitted the American-owned Colorado River Land Company (which was incorporated in Mexico) to sign a contract with Ockerson making the CRLC the entity in charge of the levee work. Despite the contract, Mexican officials on the ground refused to allow work on the river until Ockerson presented official documents. A month passed before Ockerson received the required paperwork.[50] Three weeks after construction commenced, revolutionaries in the Imperial Valley slipped across the border and seized Mexicali.

Magonista Invasion

The revolutionaries who invaded Mexicali were opponents of the aged dictator of Mexico, Porfirio Díaz. For years the Mexican government had persecuted critics, waged a merciless war against the Yaqui of Sonora, and sent opponents into exile in the United States. The Díaz administration employed an elaborate spy and assassination apparatus on both sides of the border. Revolutionaries shuttled weapons from the United States into Mexico, while plotting invasions from Texas and California. Two of Díaz's most ardent opponents, the anarcho-syndicalists brothers Enrique and Ricardo Flores Magón, in exile in Los Angeles, organized strikes and uprisings against the Mexican government.[51] In 1910, the magonistas, allied with the radical US union the Industrial Workers of the World (IWW), began preparations to capture Baja California. From their base in Holtville, the rebels plotted to seize Mexicali, a village consisting of dozens of saloons and businesses.[52]

On January 29, 1911, nineteen rebels killed the jailer and released eight insurgent prisoners as the town's customs agents, police, and residents scrambled into Calexico. Responding to frantic pleas for help, California's governor sent a small detachment of soldiers to guard Calexico, but the United States refused to send troops across the border to protect the canal system. Within a month, an army of three hundred, consisting of Mexicans, Cucapá, white and black unionists, soldiers of fortune, and American deserters, joined the insurrection. The US government asked the Díaz administration, which was fighting a rebellion in Ciudad Juárez, to send troops to guard the Alamo Canal. The governor and military chief of the northern district of Baja California, Colonel Celso Vega, marched from Ensenada to dislodge the rebels. After a battle, the rebels wounded the governor and the troops returned to Ensenada. Flush with victory, the rebels looted ranches and battled a militia of farmers and cowboys. As the rebels defended Mexicali, a renegade group of magonistas routed the garrison at Algodones, near the river, and looted and burned the customs house. Over the next few months, the insurrectionists harassed levee workers and plundered their campsite. Eventually, four hundred Mexican troops and their families landed in Ensenada. The army marched to Mexicali in April and inflicted heavy casualties on the rebels before proceeding to the river to guard the levee.[53]

A week before the levee work was completed in mid-May, the magonistas seized Tijuana, the village across the border from San Diego. Later that month, rebels forced Mexico's president, Porfirio Díaz, into exile. The new federal government's negotiators convinced the magonistas in Mexicali to surrender and accept modest financial payment to leave the region. The peace negotiators proceeded to Tijuana, but in the midst of negotiations, on 22 June, Colonel Vega routed the rebels, who fled across the border and surrendered to US troops.[54] The anarchist and Industrial Workers of the World's participation in the Baja California invasion left an indelible impression on Imperial Valley settlers, who, from this point forward, disallowed all political and economic challenges to agribusiness power. Colonizers categorized all political and economic dissenters as "agitators," "radicals," and "undesirables," who had to be subjugated like the "rebellious" Colorado River.

The investments by the Southern Pacific Railroad and the US government to control the Colorado River proved losing propositions. The Southern Pacific Railroad pressed its case for repayment for damming the river during the 1905–07 floods. After a congressional committee reduced the payment from $3 million to $773,647 in 1911, Congress again rejected repayment. In addition, the bankrupt CDC faced lawsuits and hundreds of thousands of dollars in judgments. A year after it was built, a flood destroyed most of the government's levee.[55]

Imperial Irrigation District

The magonista's threat to destroy the water system, coupled with disputes over water delivery from California Development Company's receiver W. Holabird, convinced settlers to reject private ownership of the water system. In early 1911, settlers circulated a petition requesting that the county board of supervisors call an election to determine if the voters wanted a public irrigation system to replace the CDC. In July 1911, two months after workers completed the US government's levee, voters overwhelmingly approved the creation of the Imperial Irrigation District (IID). Once the state Supreme Court ruled in favor of the IID's legal standing as an irrigation district in January 1912, the district's elected board of directors lobbied the state legislature to pass a law granting the IID powers similar to a city or county (issue bonds, levy assessments, make surveys, condemn property, and purchase private water companies). The IID was a unique institution because local voters ratified or rejected all major decisions. For instance, the water district proposed, and voters approved, a land assessment to raise money.[56]

The California Development Company's bankruptcy fractured the cooperation between it and its subsidiary La Compañía. When the railroad managed the CDC, it viewed the water distribution company as a transnational operation. But once the CDC passed from the railroad's control and into the hands of a US receiver, the CDC's responsibilities were foremost to the farmers in the Imperial Valley, and farmers in Mexico were of secondary concern. CDC receiver W. Holabird, for instance, violated the CDC's 1904 and 1910 concessions with Mexico when he refused to expand the canal system to benefit farms in Mexico. Receiver Holabird's favoritism angered La Compañía's receiver Alberto Andrade, who sued Holabird to force him to deliver water to Mexicali farmers. Andrade took great delight in threatening and harassing Holabird. On one occasion, Andrade threatened to have Holabird put in the Mexicali jail.[57]

Holabird's shift in favor of US farmers reflected a larger phenomenon in negative opinion against Mexico, and Mexicans in general, during the Mexican Revolution. Relations soured between the United States and Mexico, and on several occasions they almost went to war. The United States established a small military base outside of Calexico, and soldiers patrolled the border. The failed 1911 magonista invasion and political crises between the two nations during the Mexican Revolution sparked annexation rumors and efforts to purchase the Mexicali Valley or all of Baja California. As a result, the Mexican government stationed a military garrison in Mexicali to prevent weapons smuggling and to protect against filibusters or an attempt by the US army to seize the Alamo Canal.[58]

General unhappiness with the transnational arrangement of the water system led to talk of building a canal wholly in the United States. In 1913, an engineer commissioned by the Imperial Irrigation District estimated it would cost $12 million to build the aqueduct across the sand hills, an amount beyond the means of the settlers. Meanwhile, the IID began condemnation proceedings against the California Development Company. The IID's interest in building an aqueduct brought the Reclamation Service back into the picture. The government offered to build the waterway if the settlers agreed to put it under federal control, which was the case on the other side of the river in Yuma, Arizona. The IID refused the request because it did not have the money to pay for the project, and large landowners refused to comply with the Reclamation Service's 160-acre limitation. In stepped Mark Rose, a hard-charging rancher, who pitched a plan to Congress to build an "all-American" canal across the sand dunes to the eastern edge of the valley (called the east mesa) and finance it by selling water stock. Rose represented speculators who aimed to develop the area. Rose represented those who supported private ownership of the water system, while the IID carried the banner of public ownership of water. Rose's power play would have rendered the CDC and the IID irrelevant.[59]

Meanwhile, the fledgling Imperial Irrigation District focused on acquiring the California Development Company and its subsidiaries, La Compañía and La Sociedad. California law forbade a public utility from owning stock or land in a foreign nation, so in 1913, the IID lobbied the legislature to sponsor a constitutional amendment allowing the IID to own stock and land in Mexico. The next year California voters approved the amendment.[60] Throughout 1913, the IID and the Southern Pacific Railroad, the major bondholder and creditor of the California Development Company, negotiated a deal to sell the CDC to the IID. After a lengthy and protracted negotiation, in 1916 the railroad agreed to sell the CDC and its two Mexican subsidiaries for $3 million. The agreement required the irrigation district to pass a $3.5 million bond (the additional $500,000 for flood control) and the railroad would purchase $3 million in bonds. In an orchestrated sequence, local voters approved the bond, the CDC filed for bankruptcy, the railroad bought the CDC, and the railroad bought $3 million in Imperial Irrigation District bonds and then transferred the CDC and its subsidiaries to the IID. In the final tally, the IID acquired the CDC and its two Mexican subsidiaries, over 70,000 acres in Mexico, including the Alamo Canal (later renamed the Imperial Canal), and 200,000 acres in the United States. In a reflection of the size of the CDC's land swindle, farmers leased some 60,000 of the CDC's 200,000 acres in the United States. The IID managed the waterway and all of its landholdings in Mexico through the company's subsidiary, La Compañía.[61]

A year after the IID purchased the California Development Company, events in Mexico altered the power dynamic along the lower Colorado River. Unlike state regulation of rivers in the United States, water and land in Mexico belonged to the nation by virtue of Article 27 of the 1917 Constitution, and only citizens of Mexico (or Mexican corporations) had the legal right to use water or land. And the federal government had the power to regulate, and in some cases, to seize private property, to further the public interest. Furthermore, no foreigner could acquire direct ownership of water or property within sixty-two miles of the nation's international borders. Article 27 clearly stated that every drop of water in the Alamo Canal belonged to Mexico since the government's contract with La Compañía only permitted the right to convey water—not ownership of it. The CDC followed California law when it filed on water from the Colorado River, but the water did not flow from the riverbank to the Imperial Valley. Therefore, in reality, the CDC relinquished its legal entitlement to water from the Colorado River once it entered Mexico. Interestingly, no one challenged the CDC/IID's dubious water rights claim, for to do so would have given Mexico a stronger hand in treaty negotiations over river allocation.[62]

Political instability precluded meaningful oversight of natural resources in the Mexicali Valley until Alvaro Obregón assumed the presidency in December 1920. Soon the government stationed a federal engineer in Mexicali with the title chief of the first zone of irrigation under the Department of Agriculture and Development, to approve water-related decisions. Thus, unlike the Imperial Irrigation District, which fell under the California water code, La Compañía answered directly to the Mexican government. Irrigation engineers conducted detailed surveys of the Mexicali Valley to study its hydrology and topography for levees and irrigated agriculture.[63] Although the Colorado

River Land Company pushed back against new regulations, over time, the government asserted its authority. By the mid-1920s, foreign landowners had to obtain a permit to draw water from the Alamo Canal.[64]

All-American Canal

The shift from private to public ownership of the Alamo Canal did not quiet the voices of dissent. The thirteen privately owned mutual water companies in the Imperial Valley were saddled with debt, and settlers cautioned the Imperial Irrigation District not to purchase them. Disagreements led to the termination and resignation of IID attorneys, staff, and directors. In 1916, for example, the entire board resigned, compelling the county board of supervisors to appoint a new one.[65] Through the turmoil, the irrigation district methodically tackled each problem. At the top of the list was constructing an All-American Canal to gain total control of the water supply, as well as to stop subsidizing farmers across the border. From 1904 until 1929, the US government and American interests spent $8,354,087 for flood protection in the Mexicali Valley, while the Mexican government and Mexicali farmers contributed a paltry $621,908. Although the financial disparity is correct, what US farmers and legislators failed to mention was the California Development Company's incompetence resulting in the 1905–07 floods. Furthermore, the floods altered the stability of the riverbed, and in 1909, the river changed course to the center of the Mexicali Valley, ruining crops and pastureland.[66]

Soon after the voters created the Imperial Irrigation District, the district expressed interest in building a canal across the sand dunes to wrestle control of the river from Mexico. A lack of funds put the project off until 1917, when the district asked the Reclamation Service to survey a route. The following year a large gathering of residents agreed to pay $30,000 for a survey and $1.6 million for the right to divert the water. In 1919, local voters approved the contract with the Reclamation Service. Voters also directed the IID to negotiate a deal linking the district with a "unified Colorado River project," including a storage dam and an All-American Canal.[67] The IID also sponsored legislation requiring the Treasury Department to issue bonds to build the aqueduct. When the bill died in committee, voters sent Mark Rose to the IID board and Superior Court judge Phil Swing to Congress to get it built.[68]

The Imperial Irrigation District's bid to use federal money to build an All-American Canal held the prospect of giving settlers access to most of the river. Alarmed representatives of the Colorado River watershed (Utah, Colorado, California, Nevada, Wyoming, Arizona, and New Mexico) demanded a guaranteed share of the river, which state or federal law did not allow. The IID had tremendous leverage in the negotiations; for any Colorado River sharing deal to pass, the IID had to relinquish some of its claim to the river.[69] Representatives of the basin states met in 1919 and 1920 to hash out a plan. To streamline the talks, Congress created the Colorado River Commission. Hoping to galvanize support for the proposed aqueduct, the Reclamation Service submitted a preliminary report to Congress that recommended building the aqueduct and a storage reservoir on the river, paid for by the sale of electricity.[70]

Progress in the water negotiations emboldened settlers to purchase the thirteen private mutual water companies set up by the California Development Company to distribute water to farmers. Debt ridden, the mutual water districts were unable to keep up with the expense of cleaning the silt-laden ditches. In 1922, after the IID board passed a resolution supporting purchasing the mutual irrigation districts for $5 million, over 1,100 property owners signed a petition urging the IID to hold a bond election for $7.5 million to buy the mutuals and dig drainage canals to leech alkali from the fields. When voters overwhelmingly approved the bond sale, the IID purchased the thirteen mutual water districts.[71] The IID's budget had two major expenses: maintaining the levees in Mexico and cleaning the ditches in the Imperial Valley. The IID, however, did not consider itself an international entity; instead, it viewed itself solely as an American company, and it insisted that Mexicali farmers maintain their own ditches and pay for levee maintenance. Despite policies that favored US farmers, the IID demanded that farmers in Mexico pay the same amount for water. Mexican officials brushed aside the IID's objective for water rate equality and ordered La Compañía to charge Mexicali farmers a lower water rate and to refuse levee expenditures, which the IID used as justification to jettison Mexicali and build a canal entirely in the United States.[72]

Congressman Phil Swing and California senator Hiram Johnson introduced legislation based on the favorable report urging the government to build the All-American Canal as well as a storage dam. To promote the legislation, the Imperial Irrigation District released a three-reel film, "Putting the Collar on the Mighty Colorado River."[73] Aggressiveness by the California delegation spurred the basin states to strike a deal. In November 1922, commissioners agreed to an interstate compact, a first-of-its-kind deal among the states to carve up a river's bounty. In return for the government's promise to build a dam and aqueduct, the Imperial Irrigation District agreed to significantly reduce its claim to the river.[74] With the compact signed, but not yet ratified by the basin states, Congressman Swing and Senator Johnson spent the next six years building a coalition to pass their legislation. In 1927, for instance, Gifford Pinchot, the famed conservationist, declared the bill "the first great step toward one of the most prodigious and beneficial conquests of man over nature in the history of the world."[75] Before the bill failed that session, the House irrigation committee recommended passage to protect life and property "from the floods of the rebellious, treacherous river."[76]

Settlers earnestly desired to control the river to stop flooding and avoid droughts with a dam, but ultimately, what the Imperial Valley and the basin states wanted was to take the river away from Mexico. Every year farmers in the Mexicali Valley expanded acreage, and this meant less water for users in the United States. In congressional reports and hearings, legislators were warned: "If Mexico obtains water for its full development, it seems almost certain that a somewhat similar area in the Colorado River Basin in the United States, that otherwise would be reclaimed, will forever remain a desert." Senator Hiram Johnson appealed to the nationalistic impulses of his colleagues when he said the bill was "a measure to rescue the people of Imperial Valley, in the United States of America, from the control of Mexico."[77] Johnson frequently attacked Harry Chandler, publisher of the *Los Angeles Times* and a leading stockholder in the Colorado River Land Company (CRLC). In a speech to the prestigious Commonwealth Club of San Francisco in 1925, Hiram Johnson promised the dam and canal would rescue the "Imperial Valley from the domination" of Mexican landowners like Chandler. "The old conflict again confronts us—private greed with public welfare," he thundered.[78] In the debate over the all-American aqueduct and Boulder Dam, the CRLC served as Mexico's proxy against the legislation. Chandler supported the dam, but in congressional hearings and through his newspaper (and his proxy, the *Calexico Chronicle*), he ridiculed the proposed aqueduct to preserve his company's right to half of the water in La Compañía's canal. Chandler rallied large landowners in the Imperial Valley against the 160-acre limitation, residency requirement, and the forthcoming land assessments that would fund the project.[79]

Supporters of the proposed aqueduct habitually twisted the facts and Harry Chandler struggled mightily to persuade the electorate to oppose the new waterway. In a full-page ad in an Imperial Valley newspaper, the Colorado River Land Company rebutted the Imperial Irrigation District's financial claims by stating the district did not treat its US and Mexico customers equally. In addition to the loss of 50,000 acres of farmable land due to the floods (depriving Mexico of export customs duties and reduced employment), the IID's La Compañía did not maintain ditches in Mexico, which compelled farmers there to operate three water districts. Furthermore, the IID insisted the CRLC pay in cash before getting water, while the water bill for US farmers was often in arrears for months. Moreover, the IID subsidized US farmers when it charged them for water once it arrived in the field (a discount because of 10 percent seepage and evaporation). In contrast, the IID billed farmers in Mexico for the water as soon as it left La Compañía's canal. Finally, the IID subsidized US farmers by charging a nominal fee for water to knock down sand hills and flush alkali land, while not extending the same courtesy to its customers in Mexico.[80]

Mexico also refused to cooperate with the Imperial Irrigation District because anti-Asian nativists in California and Arizona protested efforts, beginning in 1909, to establish Japanese colonies in the Mexicali Valley, thus depriving the region of foreign investment and skilled farmers. In addition to unfounded fears of Japanese military aggression, Imperial Valley farmers objected to Japanese colonization because racists had long opposed Asian immigration and farming, and secondly, they justifiably feared Japanese farmers would reduce the Imperial Valley's share of the river. When Japanese-colonization proposals seemed imminent critics lobbied the US State Department to pressure Mexico to cancel proposals. Criticism from Mexicans against the Asians also factored into Mexico's decision to halt Japanese settlement. In October 1924, for example, *El Mercurio*, a Mexicali newspaper, reported a deal to lease 100,000 acres to 20,000 Japanese colonists. Upon hearing the news, Mexicans expressed outrage, and six Mexicali newspapers met with the governor for help in nixing the deal.[81]

Toward the final congressional debates over Boulder Dam and All-American Canal legislation, the Bureau of Reclamation commissioned Frank Adams, a professor of irrigation at the University of California, Berkeley, to conduct an examination of the historic water rights on the lower Colorado River. In his 1927 confidential report, Adams discovered La Compañía was created in August 1910 and it held a fifty-year concession from Mexico to convey water to the Mexicali Valley. Adams wrote, "Under the concession the water carried by the Mexican Company does not belong to the Mexican Company[.] It is merely a private company operating under contract for the carriage of water, title to which is vested in the Mexican government." Interestingly, someone drew a line through the last clause of the sentence! Adams proceeded to list decrees, laws, agreements, tariffs, regulations, and taxes demonstrating Mexico's ownership of the water passing into the Imperial Valley. Indisputably, the IID had no legal water rights to the Colorado River.[82]

It is unclear if legislators knew the IID had no water rights when Congress passed the Boulder Canyon Project Act in December 1928, which authorized the Bureau of Reclamation to build Boulder (later renamed Hoover) Dam, at the time the largest dam in the world. Electricity generated from the dam's power plants paid for the project. The legislation also empowered the Bureau of Reclamation to contract with the IID to build Imperial Dam, the All-American Canal, and hydroelectric power plants.[83] The government offered the Imperial Valley an incredible gift: Hoover Dam would release water during droughts and prevent flooding; it would capture millions of tons of silt that clogged ditches. Imperial Dam, which also had desilting works, served as a river's headgate for the All-American Canal. Once completed, the aqueduct would end dependence on Mexico for water. The government offered all of these goodies for free, for electricity generated from the aqueduct's hydroelectric plants would pay for the project. In return, the government expected landowners to comply with the reclamation act's provision. Once signed into law in 1929, the Department of the Interior asked California to apportion and prioritize the state's share of the Colorado River. The IID received 3.4 million acre-feet (70 percent).[84]

Opponents of the aqueduct tried frantically to convince local voters to reconsider putting the irrigation system under federal control. The IID responded, for instance, to an anticanal circular by publishing a response from a large absentee farm owner, who championed the aqueduct "because the brutal, unpalatable fact is that we in the Valley have not any 'established' and legally adjudicated water right. All agree that we have a just and sound claim to a right, but just what and how great it is has not been legally established." The remarkably candid letter continued, "No one has ever dared to venture into the stormy waters of litigation to save our water right while our claim was weighed down by the old Mexican contract and the foreign control of our main canal are like two mill-stones tied about its neck." The author believed the federal waterway would finally give the valley a "legally established water right." As an added benefit, the no-interest provision in the tentative contract would mean federal taxpayers contributed two dollars for every dollar it cost settlers for the aqueduct.[85]

After the state legislature approved the canal contract in December 1931, county residents vigorously debated the project's merits. County voters gave the IID board permission to continue negotiating a contract with the Bureau of Reclamation by a five-to-one margin in February 1932. Critics of the IID, primarily large landholders, held raucous meetings to recall the IID board or to persuade them to resign.[86] Ironically, as the supporters of the canal argued to be freed from Mexican domination, they aggressively pursued a course to subjugate the river. The Interior Department finalized the contract in December 1932, and voters spent two months pondering their decision. The contract, which noted the project was subject to the provisions of the reclamation act, required residents to pay $38.5 million in forty annual interest-free installments.[87]

Residents desperately wanted the project for it would infuse millions into the local economy during the Great Depression. Furthermore, World War I veterans who desired small farms had actively supported a federally built aqueduct. The economic downturn had caused agricultural profits to drop from $40 million annually in the late 1920s to $27 million in 1931 and $19 million in 1932. This decline lowered land values and diminished the IID's land assessment income. Mexico exacerbated the financial crunch by cutting the water rate charged to Mexicali farmers, reducing revenue from customers south of the border from $550,000 a year in the 1920s to $86,000 in 1932, compelling the IID to cut its operating expenses in half. In July 1932, the IID, for the first time, defaulted on its bonds. Under these dire economic conditions voters overwhelmingly approved the contract in January 1933. In the same election, voters approved a bond-refunding plan. With 45 percent of the district's landowners behind in land assessment payments, the IID took property title as collateral and then leased the farms back to the owners. The district also slashed the land assessment 80 percent.[88]

A major reason for the infighting was that once the voters approved the contract, the IID's irrigation supply fell under federal jurisdiction. For thirty years numerous landowners had exceeded the acreage limitation, and about half the farm owners lived outside the county. In 1930, when the executive assistant to the secretary of the interior insisted the acre limitation be included into the contract, IID's powerbrokers lobbied the Bureau of Reclamation for an exemption. Anticipating building projects where large-scale farms were commonplace, a faction within the Bureau of Reclamation pushed to ignore the 160-acre limitation. When the government finalized its contract proposal, it did not mention the acreage limitation provision.[89] The Boulder Canyon Project Act required the courts to uphold the legality of the All-American Canal project. As soon as voters approved the contract, a large farm landowner sued to overturn the contract because it did not grant an exemption to the 160-acre limitation. With the case before a superior court, the IID leaned on the Interior Department for a legal opinion exempting it from the acreage limitation. Senator Hiram Johnson and Congressman Phil Swing joined with the head of the Bureau of Reclamation to lobby for an exemption. In the last days of President Herbert Hoover's administration, the secretary of the interior signed a confidential opinion exempting the Imperial Valley from the acreage limit. Yet the IID still did not have a court decision validating its water right.[90]

Turmoil caused by the Depression, a new administration, the New Deal, legal challenges, and more pressing federal projects put the canal on hold. In May 1933, former congressman Phil Swing informed the IID directors that federal appropriations for the fiscal year were almost complete and money had not been set aside for the canal. Although the IID directors' decision to pay Phil Swing an exorbitant fee to lobby for canal funding sparked controversy, he successfully persuaded the Franklin Roosevelt administration to include money for the aqueduct among its projects.[91] Back in the Imperial Valley, the IID convinced a local superior court judge to validate the contract after furnishing the secretary of the interior's opinion exempting the 160-acre limitation.[92] In October, IID lobbyist Phil Swing met with President Franklin Roosevelt and persuaded him to authorize the first stage of the project. "Canal Battle Won!" after twenty-two years screamed a local newspaper. "End of Unemployment Here Envisioned" announced another headline. Six million dollars was allotted to begin work. The valley held massive celebrations. In Holtville, 10,000 people held a victory parade and a week of festivities. Locals

All-American Canal dedication, with Neptune granting water to the Imperial Valley, alongside white bathing beauties on raft topped with American flags, 12 October 1940. Courtesy 2Z5–9, Imperial County Historical Society, Pioneers Museum, Imperial, California.

anticipated 1,610 residents working on the canal project.[93] After considering bids, the Interior Department awarded canal contracts for the first stage in June 1934. A week later, crops withered when the river fell to its lowest point for the summer in recorded history. Potable water was trucked in as frantic farmers sent telegrams to Washington, DC for help. To funnel money into the county, the secretary of the interior released $3 million for the Imperial Dam. Honoring his twenty-two-year campaign to build the canal, Mark Rose sat atop an excavator in August and moved the first scoops of dirt. On another section, workers with horses and mules dug the aqueduct near Calexico.[94]

No sooner was the aqueduct fully operational in 1942, than skeptical officials at the Department of the Interior began to question the legality of the IID's 160-acre limitation exemption. To plead its case on this and other water matters, the IID hired a well-connected former Department of the Interior lawyer as a lobbyist.[95] Meanwhile, the IID sought revenge for being under Mexico's heel for four decades. During treaty negotiations to determine Mexico's Colorado River apportionment, the IID argued for a miniscule amount. The Roosevelt administration brushed aside the IID's petty position in order to maintain good relations with Mexico during World War II. In the historic 1944 treaty to share both the Colorado River and the Rio Grande, Mexico received 1.5 million acre-feet from the Colorado River, about twice the amount the IID wanted to give its neighbor, but less than half of what the Imperial Valley received.[96]

Conclusion

The California Development Company, the Southern Pacific Railroad, the Imperial Irrigation District, and the Bureau of Reclamation all took part in "subjugating" the Colorado River. Efforts to divert and control one of the continent's mightiest rivers, settle and irrigate the nation's most forbidding desert, while ignoring a treaty with Mexico, circumventing homestead laws, and lobbying to get an exemption from the 160-acre limitation, reveal the audacity, determination, greed, and dark ambition of the colonizers. A combination of arrogance, recklessness, and incompetence resulted in a series of floods between 1905 and 1907, requiring a herculean effort by the Southern Pacific Railroad to return the river to its channel. Earlier, in 1904, Mexico granted the California Development Company's subsidiary, La Sociedad a concession to convey water to the Mexicali Valley. Unable to keep up with escalating operating expenses, in 1909, the CDC and its Mexico subsidiary La Sociedad went into receivership. A year later, the CDC created La Compañía to operate the Alamo Canal.

Private enterprise had failed, and the settlers turned to public ownership of the water system, creating the Imperial Irrigation District in 1911. The IID's improbable success at turning around the California Development Company's colossal engineering failure was due to shrewd, tenacious leaders, public ownership of the water system, governmental assistance, a media blitz promoting irrigated agriculture and the subjugation of nature, and sheer luck. The Imperial Valley flowered when the electorate made decisions. Voters approved purchasing the California Development Company and its two subsidiaries in 1916, and the thirteen mutual water companies in 1922. Voters also authorized the Bureau of Reclamation contract to build the All-American Canal, Imperial Dam, and hydroelectric power plants on the aqueduct to free the Imperial Valley from dependence on Mexico for its water. The basin states and the federal government were so determined to wrestle the river from Mexico they did not challenge the California Development Company's dubious water rights claim. The evidence presented here argues the CDC, and its successor, the Imperial Irrigation District, forfeited its legal entitlement to the river under California law once the water flowed into Mexico. Under this scenario, the federal government should hold the valley's water rights since it built and has never relinquished title to the All-American Canal.

Nevertheless, the IID was instrumental in persuading Congress to pass the 1928 Boulder Canyon Project Act, which authorized the building of Hoover

Dam, a major step for supremacy over the river. The legislation elicited praise and trepidation: "They walled the giant out with concrete gates and rock riveted levees," wrote a 1930 local history book, "but to this very day there exists the justified fear that the giant merely sleeps. The conquest is still on and not until the walls of Boulder Canyon are sealed and the giant put in prison behind them will that feeling entirely disappear."[97] When the dam was completed in 1935, the government called attention to its accomplishment: "THE ONCE WILD Colorado River has been tamed," announced a Bureau of Reclamation pamphlet. "Completion of Hoover Dam" has "closed an initial chapter in man's long effort to harness the resources of one of the world's most treacherous rivers."[98]

Boulder Dam ended flooding and gave the region downstream an even flow of water, while the aqueduct quenched the thirst of the Imperial Valley's agribusiness empire. Colonizers had finally whipped their two major nemeses—the river and Mexico. Yet the Imperial Valley had a complex relationship with the federal government. With one fist clenched and another hand extended out for help, the Imperial Valley, from its conception, was a contradiction. Settlers tacked back and forth between begging for government assistance in times of crisis and demanding local control otherwise. This pattern became evident when the two parties clashed over immigration policies and labor relations issues.

Fear of the river's destructive power laid the foundation for a psychology of control that led to a manipulation of all aspects of nature. Settlers reasoned that if the river could be subjugated and made to produce crops, then farmers could force their will on all aspects of the environment. Colonization, engineering the landscape for industrial farming, and ridding the landscape of nature's "pests" are explored in the next chapter.

CHAPTER 2

Colonizing the Desert

> The conqueror of the desert is at hand. Look down, now, on that great expanse of burning waste for the last time. [Nature has waited for] the touch of man's hand to blossom into a garden of incomparable beauty and richness.
>
> Edgar Howe and Wilbur Hall,
> *Story of the First Decade in Imperial Valley, California*

Madge Morris Wagner published a poem in 1911 celebrating the colonization of the Imperial Valley. Its sensual language feminized nature while expressing the dark side of her awesome power.

> To the Colorado Desert
> Thou brown, bare-breasted, voiceless mystery,
> Hot sphinx of nature, cactus-crowned, what hast though done?
> Unclothed and mute as when the groans of chaos turned
> Thy naked, burning bosom to the sun. . . .
> And thirsty mouths pressed on the sweltering sands,
> Mark here and there a gruesome, graveless spot,
> Where someone drank thy scorching hotness, and is not.
> God must have made thee in his anger, and forgot[1]

Wagner's poem sheds insight into the mindset of the settlers as they transformed the desert into farms. In its virgin form, the inhospitable landscape, recalled a "pioneer," "waited, silent and hot and fierce in its desolation, holding its treasures under the seal of death against the coming of the strong ones" who were determined to conquer it.[2]

Colonizers engineered this patch of arid America into an irrigated agricultural powerhouse. The bulldog-like effort to control the Colorado River was the first step in transforming the ecosystem and terrain for commercial agriculture. This chapter continues the theme of environmental subjugation by examining colonization of the landscape as well as the establishment of a racially segregated, rigid, class-based social structure. Boosters and residents promoted a myth of egalitarian settlement for three main reasons: to validate their hard work and sacrifice; to lure prospective settlers to the desert; and, finally, to garner government assistance. *The Winning of Barbara Worth* (1911), by Harold Bell Wright, was the principal text validating the "collaring" of the river and the promotion of white yeoman farmers. The novel's themes are encapsulated in a love story: the "winning" of the affections of Barbara Worth, a young white woman. Wright's generation promoted desert reclamation, a secular form of spiritual regeneration. Just as cultural reclamation meant "civilizing" Native Americans, reclamation of the landscape transformed "worthless" land into productive fields.

Race and gender profoundly impacted the creation and perpetuation of the valley's power structure. Authority flowed from landownership and racial privileges. Leaders of the California Development Company envisioned a community dominated by white landowners. All the investors, politicians, homesteaders, and all of the farm landowners were white. Irrigation system officials and farmers, however, hired Indians and Mexicans, laying the foundation for a multiracial society of predominantly adult males.

Settlers had firm ideas about individual and collective social behavior. As the first chapter suggests, the psychological terror of the Colorado River and the threat by magonistas to destroy the Alamo Canal terrified the settlers and resulted in an intense desire to control all aspects of life. From that point forward, the conservative Republican agribusiness elite brokered no challenge to its authority. As a result, community leaders engaged in extreme repression to support a regime of social control. Political radicals and social deviants were thrown in jail, beaten by vigilantes, and banished from the county or deported. Subduing the landscape and eradicating nature's "pests" encompassed another component of colonization. Settlers leveled the desert and flooded the soil to eliminate alkali. Lacking experience with desert irrigation, farmers appealed to federal and state farm experts for technical and financial support to exterminate plant diseases, bugs, coyotes, birds, and rodents.

Colonizing Land

In 1900 the California Development Company (CDC) created the Imperial Land Company to advertise, recruit settlers and investors, facilitate homesteading, and sell water stock. Imperial Land was both a colonizing agency and a holding company for the mutual water companies' water stock, which received 25 percent of the gross sales of water stock. It had the rights to survey, plat, and sell town property and was invested with all rights to power, light, telephone, and railroad rights-of-way.[3] Imperial Land marked town sites, contracting with speculators and realty companies to develop urban property. Realtors sponsored railroad excursions for newspaper reporters, bankers, and prospective buyers. The railroad carried newspapers to Los Angeles daily, supplying potential buyers with information on land prices and desert living conditions.[4] Booster pamphlets and newspaper advertisements in midwestern newspapers promised a new desert Eden. One surviving tract read: "Two years ago this Colorado Delta was a trackless waste of worthless land—all the property of the Government—and was not considered worth 30 cents a section."[5]

Realtors took prospective settlers to the US Land Office in Los Angeles to pick parcels. Citizens filed petitions for land under the Homestead Act or the Desert Land Act. Most settlers obtained land under the Desert Land Act because the Homestead Act had a six-month residency requirement. With 25 cents an acre down, Desert Land Act homesteaders could buy 40 to 320 acres. Within three years, they had to show improvements equaling one dollar an acre for three years (water stock purchased from Imperial Land Company met these requirements). The total investment was $3.25 an acre payable at the end of three years. Significantly, residence on the land was not required.[6]

Imperial Valley's land boom lured legitimate homesteaders and crooks looking to turn a quick profit. Land agents acquired property illegally when they hired individuals to file on a 320-acre desert claim and then transfer title to Imperial Land Company. Since the Desert Land Act did not require residency—only land improvements—many "homesteaders" were actually investors who made a quick visit to the desert, bought water stock, and never returned. Before water gushed into the Imperial Valley in 1901, land filers held over 70,000 acres. The following year, a total of 125,000 acres were claimed. By 1904, an estimated 9,000 souls resided in the desert, on farms totaling 150,000 acres. Four years later, homesteaders had taken all the available land. Unimproved land with access to canals sold for $20 to $75 per acre, based on location and proximity to the railroad.[7] The California Development Company cooperated with the Southern Pacific Railroad to develop the region. The railroad anticipated earning money shipping passengers, crops, and freight. As the county's largest landowner, the railroad aimed to sell its property on the northeastern edge of the basin. To encourage settlement, it donated $500 for schools and it built rail lines to the towns. The railroad's investments and objectives influenced Edward Harriman, the owner of the Southern Pacific, to take over the CDC in 1905 to prevent flooding. The railroad also lent equipment and provided legal advice to the mutual water companies.[8] Speculators made a windfall in the murky world of land scrip, water stock, phantom homestead applicants, and foreclosed homesteads. Los Angeles, San Francisco, and Portland capitalists purchased enormous tracts on both sides of the border. By 1919, eight landholders owned over 2,000 acres. California Development Company engineer Charles Rockwood, for instance, held title to 9,000 acres, and a Los Angeles banker got his hands on 4,600 acres before selling it for $12 an acre.[9]

C & M Ranch and the Colorado River Land Company

A syndicate of Los Angeles investors owned the largest fiefdom on both sides of the border. Harrison Gray Otis, the longtime publisher of the *Los Angeles Times*, and Moses Sherman were major stockholders of the binational operation. When Otis passed away, Harry Chandler, his son-in-law, became the publisher of the *Los Angeles Times* and took a leadership role in the syndicate's land investments in the Imperial-Mexicali Valley. In 1902, the Otis-led investors created the California-Mexico Land and Cattle Company (C & M Ranch). The ranch owners used an assortment of fictitious Desert Land Act homesteaders and land purchases to cobble together a large estate adjacent to the border. The C & M Ranch became a model for scientific farming and ranching. That same year, the syndicate formed the Colorado River Land Company, S.A. (CRLC), which over the next few years bought most of the Mexicali Valley, 832,337 acres, from Guillermo Andrade and various Americans.[10]

An early history of the valley summed up the awe-inspiring transnational operation: "The venture was on a scale that exceeded in size anything of its kind in the United States." It was a cattle ranch and a cotton-making factory. "The initial development of the ranch on the American side made it the show place of the Valley and 15,000 acres planted on the Mexican side produced magnificent crops year after year." C & M's success "was a mighty factor in adding to the reputation of the Valley as a producing area second to none." In 1912 the Mexicali-based CRLC shifted from cattle ranching to cotton farming with Chinese tenants and transnational Chinese migrants. By 1920, the CRLC proclaimed itself the largest cotton ranch in the world. Wary of nationalist sentiment against large foreign landowners and determined to curb Mexican settlement, the CRLC rarely leased or sold land to Mexicans. When California forbade Asian ownership of farmland, the CRLC leased ground to Japanese and South Asian tenants.[11]

Linking Water to Land

When engineer George Chaffey took over the California Development Company he did not plan to get rich owning land or selling water. He anticipated becoming wealthy selling water stock. Chaffey's master plan began with the formation of mutual water companies who bought water from the CDC and distributed it to specific areas of the Imperial Valley. Complicating matters for the engineers designing the gravity irrigation distribution system, the 1905–07 floods created two canyons on opposite sides of the valley known as the Alamo River on the eastern edge and the New River on the western side. Filled with field water runoff, these two streams flowed northward to the Salton Sea.

Chaffey set up the mutual water companies to operate at cost. Farmers paid

Mules smoothing the edge of the New River Canyon, ca. 1910. Courtesy Inventory of the James D. Schuyler Papers, bulk 1886–1912, Box 2, folder 13, item no. 71, From the Holdings of the Special Collections and Archives, Water Resources Collections, UCR Libraries, University of California, Riverside.

the mutuals $1 for every acre-foot of water (enough to cover one acre with a foot of water) they ordered. The mutual water companies split each dollar in half, paying the California Development Company 50 cents per acre-foot from the Alamo Canal and keeping the rest for operating expenses—namely, for clearing the canals of weeds and silt. Imperial Land Company realtors walked landowners through the water stock maze. From the Los Angeles Land Office, agents whisked homesteaders to see property and to buy water stock. For the right to purchase water from the mutual water companies, landowners had to buy one share of water stock for each acre. The price of water stock was tied to the value of an acre, ranging from $10 to $15.[12] At the beginning of the land rush, Imperial Land peddled 50,000 shares of water stock for $10 a share, which artificially inflated land values. Eager to attract settlers, each share sold for a $1 down payment and $1 yearly payments. For those landowners who could not afford water stock, Imperial Land kept the property title as security and settlers made payments with 6 percent interest. Desperate for cash, labor, and supplies to dig the water system, the CDC and Imperial Land traded water stock for labor on the canals and for mules and wagons. Since water stock equaled land scrip (federally issued bonds) in value,

Imperial Land allowed the mutuals to trade water stock for land scrip for as little as 10 cents on the dollar.[13]

Chaffey's master plan enriched Imperial Land Company stockholders for when settlers went broke, the company foreclosed and sold or leased the property. Through these machinations, the CDC controlled settlement. By separating the CDC's water system from Imperial Land, observed historian David Lavender, Chaffey and associates "could not only absorb land without creditors immediately attaching it, but they could also resell later on without accounting for the proceeds."[14] When the Southern Pacific Railroad controlled the California Development Company from 1905 to 1909, the railroad expanded Chaffey's enterprise of turning water stock into capital. Under railroad management, the CDC and the mutual water companies sharply increased water rates to pay for spiraling costs.[15] Homesteaders and speculators ran into a buzz saw of water stock fees, intolerable weather, debt, and flooding. Flooding and the spike in water fees led to water stock foreclosures and the California Development Company's acquisition of more property. These shenanigans explain why the CDC owned 200,000 acres, almost a third of the valley's farmable land, and leased 60,000 acres.[16] Within the first decade, the gravity distribution system consisted of a dozen intermediate canals measuring 117 miles. Off the intermediate canals jutted 179 lateral ditches totaling nine hundred miles. By 1920, the thirteen mutual water companies operated almost three thousand miles of canals.[17]

Establishing the Structure of Power

Power and authority stemmed from the melding of interests among landowning farmers, absentee owners, tenants, businessmen, and public officials in support of irrigated agribusiness. Settlers recorded this history in hagiographic texts celebrating colonization. For instance, newspaper editor Otis Tout's *The First Thirty Years* tells a glorious tale of nature's subjugation and includes vignettes of successful male "pioneers." Upon closer examination, the book describes rampant speculation, absentee ownership, and interrelationships of power, aptly portrayed in the following examples.

Myron Witter arrived during the floods and acquired a homestead. A newspaperman with large ambitions, he worked for a year at a newspaper before moving on to the *Brawley News*, which he bought and ran for a quarter of a century. In 1924, Witter parlayed his support for farming into a seat in the state Assembly.[18] What his biography failed to mention is, through his newspaper Witter defended agribusiness against a litany of critics. As an absentee farm owner, Witter chronicled agricultural and water issues while minimizing the hardships of farm laborers and cheering for repressive measures against unionized farm laborers. In addition, absentee landowning lawyers provided instrumental support to irrigated farming. Marvin Conkling left a Los Angeles law firm in 1906 to help establish the town of El Centro as the county seat. He drew up the documents enabling the Imperial Valley to cede from San Diego County. Over the next two decades he served as El Centro's attorney, helped amend the law to create the Imperial Irrigation District (IID), and served as the water district's counsel before becoming a judge. Conkling's biography neglected to mention the Japanese tenants who leased his 400-acre farm.[19]

Two more examples show how businessmen facilitated successful desert farming. The brothers William and Leroy Holt, the valley's primary financiers and businessmen, illustrate the speculative nature of colonization and the interwoven structure of power. Experienced investors and bankers, the Holts took jobs in the first bank in Imperial. After purchasing the bank, Leroy went on to start other banks, eventually becoming the county's most important lender. William snapped up discounted water stock, eventually acquiring 18,000 shares (equal to 18,000 acres) for $50,000. In 1903, he established Holtville on the east mesa. He bought 640 acres nearby and used government land scrip to buy property for his new town. By 1912, William created a colonization company to develop the area. Because the land sat high up on the mesa, he and rancher Mark Rose created two irrigation companies to water the area. The quest to get water to the east-side mesa fueled the impetus to build the All-American Canal.[20]

The Holts are worth considering in detail. They wielded financial power because the federal farm bank refused to lend money to settlers for fear of Colorado River flooding. Their ownership and management of banks gave them intimate knowledge of the financial solvency of farmers, and they ingratiated themselves with landowners who leased land to tenants. Leroy persuaded the Bank of Italy to establish branches in the valley. He crowned his banking career as the vice president and regional manager of the Bank of Italy (later renamed the Bank of America) where he dispensed loans and handled foreclosures, facilitating the transfer of tens of thousands of acres to the bank. William owned newspapers, banks, a telephone company, a church, hotels, and other businesses. In 1904, William and a partner bought 320 acres to create the town of El Centro, which became the county seat and the region's commercial and banking center.[21]

The Holts expanded their financial power and landholdings into influence over water decisions. William Holt represented the Imperial Valley Water Users in 1904 in negotiations with the Reclamation Service to purchase the California Development Company. His brother, Leroy, served as treasurer of mutual irrigation companies. Leroy later became the Imperial Irrigation District's principal banker during its infancy. When the strife-ridden IID board resigned en masse, in 1916, the county board of supervisors appointed five replacements, with Leroy Holt chairing the new board.[22] The Holts and like-minded capitalists, large landowners and lawyers, controlled the California Development Company, the mutual water companies, the Imperial Irrigation District, and the county board of supervisors. These agribusiness elites handpicked and directed judges and police to do their bidding, all in the quest for commercial farming domination over nature and fieldworkers.

The Egalitarian Myth of Colonization

In stark contrast to the rampant speculation and consolidation of vast estates, settlers and boosters promoted colonization as a grand egalitarian effort. They pointed to homesteading laws and the mutual irrigation districts as proof of an egalitarian society. Promoters pointed out that landowners owned the mutual water districts through water stock ownership, and each share of water stock entitled the landowner to one vote in district elections—one acre, one vote. The one acre one vote model was viewed as a way to protect against speculators and to support family farms.[23] Irrigation district laws, which gained popularity in the 1870s and 1880s, were intended "to promote small scale, diversified agriculture and democratic values," argues historian Donald Pisani. He notes, however, the intentions of the homestead acts and the irrigation districts "had by the 1920s and 1930s been perverted to serve as the foundation for agribusiness."[24] Thus when boosters promoted a Jeffersonian vision of yeoman farming, they meant the equal opportunity to file on land, own water stock, and vote in proportion to their landholdings.

The Winning of Barbara Worth (1911) fictionalized the ideal of the yeoman democratic farmer conquering the desert for irrigation agriculture. The sweeping epic sprang from the pen of former preacher and artist turned author, Imperial Valley resident Harold Bell Wright. Published one decade after George Chaffey diverted water to the region, Wright recycled and updated the Jeffersonian myth. Farmers, joined by reclamation-minded capitalists, marched hand in hand to answer William Smythe's clarion call to settle the arid West with irrigated farms. He told a grand tale refuting the negative press about the floods and celebrating the valley's magical transformation. Combining Jeffersonian ideals with an old-fashioned love story, the plot carries the reader through hope, tragedy, and rebirth through water.[25]

Wright's description of nature, white settlers, Indians, and Mexicans illustrated the viewpoint of the colonizers. The desert was forbidding and deadly: "To the land of lean want, of gray death, of gaunt hunger, and torturing thirst" came the water masters. When Jefferson Worth, the reclamation financier, looked out over the desert, he imagined its transformation. "He saw first, following the nearly finished work of the engineers, an army of men beginning at the river and pushing out into the desert with their canals, bringing with them the life-giving water." Once white settlers arrived: "Hummocks would be leveled, washes and arroyos filled, ditches would" carry water to grow "great fields of luxuriant alfalfa, billowing acres of grain, with miles upon miles of orchards, vineyards and groves." In the final act to this romantic tale: "The fierce desert life would give way to the herds and flocks and the home life of the farmer." Jefferson Worth scanned the horizon and thought: "The wealth of an empire lay in that land under the banker's eye, and Capital held the key" to unlock nature's boundless riches. Yet before profits could be wrung from the earth, settlers had to break nature's death grip: "Steadily mile by mile, acre by acre, and at times almost foot by foot, the pioneers wrestled their future farms and homes from the dreadful forces that had held them for ages."[26]

Harold Bell Wright's characters embodied commonly held racial stereotypes. All the major characters were white. Racial minorities played bit parts and were happy to serve the desires of white heroes. Wright portrays Mexicans and Indians as either hard-working laborers and servants or robbers. In Jefferson Worth's vision quoted above, engineers, canal diggers, and farmers are all white. Elsewhere, Indians are happy to work for whites and Indians bow when they approach the Worth family. Indians are innocent custodians of "worthless" land who unwittingly cooperated in the conquest of the environment. José, an old Indian, for instance, unlocks the key to the desert by showing the Seer the path to divert the river.[27] This event actually occurred when a Cucapá showed George Chaffey the best path for the Alamo Canal.

The Winning of Barbara Worth's powerful message celebrated untrammeled individualism and unfettered capitalism. Bureaucrats are nowhere to be found. Private capital and individual initiative take the river and win. The original manuscript and photos of Wright's home and family are on display in the Imperial Valley's Pioneers Museum. In the exhibit is a letter from Ronald Reagan praising the book. Reagan wrote that when he read *The Winning of Barbara Worth* in his youth it had a powerful philosophical impact. Reagan adored the message of unbridled individualism. The book, a five-hundred-page blockbuster, was adapted to the stage, and the 1926 film version is one of the most popular silent movies ever made.[28] The story's images cemented the valley as a model of white, Christian, virtuous, democratic citizenry, taking up God's command to go forth and tame the wilderness.

Colonization after the Floods

In the wake of the flood's devastation, settlers scanned the region and saw a host of problems, but also vast possibilities. Land titles remained unpatented, the Southern Pacific Railroad managed the California Development Company, the river seemed poised to flood the region, farmers could not get credit, and the mutual water companies lacked capital. First, in 1907, settlers voted to cleave 4,597 square miles from the eastern half of San Diego County and name it Imperial County. Next, the secretary of the interior ordered surveyors to correct the maps of the international boundary. In 1908, the government approved new border maps and patented homesteads. Soon rumors of incorrect border surveys reached the government. After another investigation, the Interior Department concluded the surveyors had committed fraud, requiring yet another survey. When the redrawn maps were completed, some people lost homesteads or had property reduced.[29] The confusion over the adjusted survey boundaries attracted swindlers who fleeced unsuspecting greenhorns. There were two groups of crooks, agents of Imperial Land and land hustling "sharpers" who investigated parcels with unpaid taxes or found violations of the land grant terms in order to challenge them in court.[30] Settler historians castigated "land sharpers" for preying on unsuspecting settlers. Sharpers investigated "titles very closely and, aided by the inconsistencies and blunders of land office officials, they found many ranches where technical errors had been made either in filing, purchasing from original entrymen, or making proof of work done."[31]

After investigating land grant irregularities, the government prosecuted flagrant violators. In 1908, the US district attorney announced a conspiracy by Imperial Land to defraud the Land Bureau. Six of the eight men under investigation lived on the coast. The indictment claimed they hired "dummy entry men" to file on land, cover it with water stock, and then sell it. One pled guilty and paid an eye-popping $5,000 fine, two paid $250 each, and another was discharged. After a long trial one was released and two brothers named Chaplin had to pay $1,000 each and serve nine-month jail sentences, which they appealed.[32] According to an observer: "A very large number of the best ranches in the Valley are now held by entrymen who took them on assignment from other persons who had made it a business to file on land, cover it with water stock (purchased 10 cents on the dollar) and sell it, repeating this operation several times."[33] In 1912, the Chaplins lost their appeal. A newspaper commented: "Considerable sympathy is expressed in the valley for the Chaplins, for though guilty of a technical violation of the desert land law, this law was loosely interpreted in the early days" of pioneering.[34] The newspaper's reporting of the Chaplins' fraud is an early example of how the press minimized the criminal behavior of land agents and their clients. These swindles contradict the laudatory celebrations of the conquest of the desert by morally upright settlers. The "founders" excused Imperial Land for committing fraud, yet took umbrage with outside defrauders. The whitewashing of history—excusing away gross incompetence and criminal behavior—became a pattern in the local newspapers and in the celebratory histories local historians wrote.

Bitter quarrels between federal authorities and settlers laid the foundation for a schizophrenic relationship. Control and dependency went hand in glove.

On one hand, government officials were impressed by the settlers' ambition and sense of mission to carve a Garden of Eden out the forbidding desert. Many officials gladly dispensed government support to subjugate nature and fulfill the nationalistic mission of settling the West. Other bureaucrats were appalled by the lack of respect for the government and the flaunting of laws. Settlers also held contradictory views toward the government. They yearned for local control, but were dependent on government assistance.[35]

After the initial rush of homesteading, the east and west mesas above the valley floor remained outside the reach of the gravity irrigation system, leaving the north end of the basin south of the Salton Sea as the only place to expand. Much of this land was owned by the railroad. Speculators bought large tracts from the railroad and lured settlers. One syndicate, the Imperial Valley Farm Lands Association, led by Harry Chandler, owner of the *Los Angeles Times*, bought 47,000 acres from the railroad for $1.6 million in 1912. Eight years later, the Chandler syndicate sold a tract of 37,000 acres. Behemoths like Balfour-Guthrie, a British cattle firm, purchased 10,000 acres for cotton and cotton by-products for cattle feed. Another syndicate acquired 10,000 acres to grow cotton.[36] Corporate landholders hired tenants, built canals, and demanded service from the Imperial Irrigation District. These fiefdoms account for the explosion of cultivated land, from 225,000 acres to 350,000 acres ten years later.[37]

During World War I, the demand for dairy products, livestock feed, cotton, beef, and vegetables prompted farming interests to pressure the Interior Department to open the east mesa to homesteading. Local leaders saw World War I veterans as prime settlers. The expansion of farm acreage, however, was constrained by the Interior Department's policy of restricting homesteading to areas supplied by irrigation. Although the east mesa sat higher than the canal, settlers lobbied to open up this portion of the public domain and to give preference to veterans. The Land Office reluctantly accepted applications for homesteads on 12,000 acres.[38] Three months later, little land remained because "squatters" claimed preference by right of residence. Bands of roaming, penniless prospective homesteaders wandered the desert west of the Salton Sea. The push to open the public domain further succeeded. A 1920 Land Office report noted that "the government has approved filings of homestead and desert entry claims for 180,000 acres of land, meaning approximately 10,000 people have profited by taking advantage of the granting to citizens of the public domain."[39] A newspaper story, under the headline "Plucky Woman Conquers Land," reported the difficulties "reclaiming" desert land. Niland homesteader Mrs. Al Martin had operated a farm for three years. "Alone, but fearless, and with her trusty rifle, her favorite greyhound, her cat family and then the chickens and turkeys, she has found solace and comfort out on the very edge of the desert." Martin grew "ten acres of corn, has a patch of alfalfa, has raised barley, does her own irrigating and looks after the situation much better than most men would."[40]

In the early 1920s, World War I veterans relocated to the Imperial Valley in a program called the "One-Man Farm," sponsored by the Allied Chambers of Commerce. The Veterans Bureau provided loans to buy small plots of twenty to forty acres and assisted settlers. Absentee owners were pressured to sell land at reasonable prices. The Farm Bureau, horticultural commissioner, and state agricultural officials held classes and provided technical advice in irrigated desert farming. The back-to-the-land movement merged with the campaign to replace Japanese and South Asian tenants with white conservative farmers to buttress against "un-American" nationalities. The effort was also a clever way to wrestle land away from absentee landowners. The resettlement of veterans profoundly impacted employer-labor relations over the next two decades. Under the newly formed American Legion and the Farm Bureau, veterans served as sentinels, reporting crimes, policing alleged immorality, and operating as vigilantes during strikes.[41] Advocates of white supremacy encouraged veterans to replace Asian tenants. Several years after the passage of the 1920 California Alien Land Law banning Asian farm leases and Asian corporate farm ownership, an editorial in the local newspaper congratulated the Brawley High School for developing "the plan of instruction now adapted for the entire United States trainee agricultural system" to provide scientific instruction to veterans. The high school program was so successful landowners hired one hundred veterans "of the best type to this community, to turn the land from cotton patches to gardens, orchards and vineyards, a home on every 20 acres, with red-blooded Americans, instead of turban topped Hindus."[42]

Social Demographics

The Homestead and Desert Land Act's stipulation that only citizens or those petitioning for naturalization could file on public land resulted in a mostly white landowning society. The 1910 census counted 13,591 residents in Imperial County and adult American-born whites predominated. There were twice as many males (8,900) as females (4,691), and almost half of the residents were over age twenty-one. Census enumerators counted a thousand European immigrants from nineteen nations. People of color consisted of 1,461 Mexicans, 217 Japanese, sixty-five blacks, thirty-two Chinese, and thirteen others. There were also 682 Native Americans. The government and the courts classified Mexicans as Caucasians, or racially white, meaning Mexicans could not be legally segregated in California, and they were not subject to antimiscegenation and discriminatory laws affecting Asians and Native Americans. In reality, whites treated ethnic Mexicans as people of color, and they endured discrimination and segregation along with others.[43]

In 1920, the census revealed a dramatic shift in the county's social demographics. From 1910 to 1920 the population tripled from 13,591 to 43,453, with people of color comprising a quarter of the population, along with 2,500

European immigrants. The male (25,966) to female (17,487) ratio significantly narrowed. The Mexican population rose from 1,461 in 1910 to 6,414. Japanese increased from 217 to 1,986. Although the number of Indians expanded from 682 to 1,185, most lived on the Indian reservation on the eastern edge of the county. Black residents skyrocketed from sixty-five to 1,648. Chinese increased from thirty-two to eighty-eight. Finally, the "Other" category (South Asians, Filipinos, and Koreans) increased from thirteen to 381.[44] A host of factors explain the influx of blacks and immigrants. Isolation, weather, and the difficulty adjusting to irrigated desert farming persuaded many homesteaders to move away and hire Asian and black tenants. Meanwhile, a scarcity of labor prompted the recruitment of southern blacks and foreigners.[45] In 1910, most of the 13,591 residents lived on farms and in a few small hamlets. The largest towns were El Centro's 1,610 residents, followed by Imperial with 1,257. Calexico, Holtville, and Brawley each had less than nine hundred people. A decade later, most of the county's 43,453 inhabitants still lived in the rural hinterlands. The three largest towns were Calexico, whose 6,223 residents served the border economy; El Centro had 5,464 souls; and Brawley had 5,389 residents. Ethnic and racial groups self-segregated, forming separate commercial, civic, and social organizations for mutual benefit and for self-protection.[46]

On the other side of the international border, the Colorado River Land Company's farming and labor policies profoundly shaped settlement patterns. The town of Mexicali, across the line from Calexico, was then, and remains today, the only urban entity in the Mexicali Valley. Originally, Mexicali existed as a customs and immigration port of entry. After the floods, the village consisted of low-slung saloons and brush and adobe homes for workers.[47] In December 1914, Colonel Esteban Cantú, the military commander of the Mexicali garrison, unseated the unpopular governor and declared himself governor (1915–20). Cantú moved the capital from the coastal town of Ensenada to Mexicali, where he declared his neutrality during the Mexican Revolution. To maintain his independence, and to pay his troops and operate the government, Cantú instituted policies to grow the cotton industry. Following the Colorado River Land Company's suggestion, Cantú encouraged victims of anti-Chinese riots in other parts of the country to relocate to Mexicali and earn a living as tenants, laborers, and merchants. The governor also allowed farmers to import Chinese transnational workers, while charging a $130 to $145 fee for each person, a tax the governor split between the government and himself. The Colorado River Land Company gladly accepted Cantú's labor support, while complaining about his high land taxes.[48]

Governor Cantú also enriched himself and funded the government by attracting American tourists chafing under laws outlawing drinking and gambling. He aggressively expanded Mexicali's vice industries by licensing ornate gambling and drinking establishments, dance halls, and restaurants. The gov-

Governor Colonel Esteban Cantú (first row, seventh from the left), military officers and dignitaries in the Mexicali military fortress, celebrating the opening of Cantú's highway to the coast, 1919. Courtesy P1999_104_035, Imperial County Historical Society, Pioneers Museum, Imperial, California.

ernor also licensed hundreds of prostitutes from around the world and sold licenses to opium dealers. Cantú's tourism and profarming policies reaped a windfall in taxes and a fortune in bribes. The governor used the brimming treasury to modernize the district capital. Cantú also spent tax revenue to enlarge his well-paid military force and billet soldiers in a new fortress. Under Cantú's pax Mexicana, the population of the town of Mexicali grew from 462 in 1910 to 6,782 in 1921 to 14,842 in 1930. For an indication of the international dimension of Mexicali, the 1921 census counted about 3,000 Chinese, 400 Japanese, and 1,400 Americans. The population in the hinterland equaled the number in the town. Esteban Cantú wielded absolute power in the region until he refused to submit to the authority of incoming president Alvaro Obregón, who sent an expedition headed by General Abelardo Rodríguez, the northern district's former military chief (and a future governor of the district and a future president of Mexico), to oust Cantú in 1920.[49]

The Etiquette of Power

Settlers focused on the subjugation of nature, the perfection of agribusiness, procuring the ideal labor force, and establishing a society predicated on morality and white supremacy. Elites used a variety of instruments of control to enforce this vision, including morality statutes, racial segregation, and ridding the towns of "loafers," assorted social malcontents and criminals. The structure of white supremacy and social control rested on a foundation of legal and political hegemony. From its inception, conservative Republican agribusiness elites ran the valley. Politicians, police, judges, and businessmen were frequently farm or ranch owners. The interlocking nature of political and economic power policed individual and group behavior. To maintain social control of the working class, local leaders outlawed alcohol, drugs, prostitution, and gambling, while looking the other way when members of the elite partook in illegal activity. Community leaders assaulted, intimidated, jailed, deported, or banished social, political, and labor dissenters. Farm elites knew interracial activity blurred social boundaries, and they vigorously enforced racial separation. Segregation reinforced white hegemony and disempowered racial minorities. Critics could not express grievances or alter the political and legal system. Immigrants of color could not vote because they were ineligible for citizenship, while very few Mexicans became citizens.[50]

The trajectories of power frequently shifted as individuals resisted or mediated the imposition of authority. Enforcing a moral, law-abiding, segregated community, while also policing underpaid mobile workers, proved extremely difficult. Immigrants of color and marginalized whites used a plethora of resistance strategies, ranging from abandoning abusive employers, physical confrontations, theft, arson, and appeals to the Mexican government for redress during strikes. Despite intense ethnic and racial distrust and violence, over time, people of color and working-class whites developed friendships, business partnerships, sexual relationships, and labor alliances.[51]

Colonizers envisioned an agribusiness utopia resting on Jeffersonian egalitarian ideals and Progressive-era Anglo middle-class social mores that emphasized order, efficiency, sobriety, hygiene, and racial separation. White leaders enacted and enforced morality statutes to control "immoral" behavior, maintain racial boundaries, and shape working-class behavior. Imperial County's first ordinance prohibited liquor consumption and the third forbade gambling. Local, state, and federal laws outlawed prostitution, interracial marriage, and narcotics.[52] Leaders kept a tight lid on vice until 1912, when the town of Imperial permitted liquor consumption. Overnight, saloons, prostitutes, drug pushers, and gamblers appeared in different towns. In one instance, Brawley police surprised thirteen blacks shooting craps and fifteen Mexicans playing cards. "The darkies and Mexicans were lined up and relieved of all available cash," reported the local newspaper. Corrupt police and officials undermined sobriety. In 1914, Imperial's marshal was arrested for receiving bribes, and the following year, critics claimed Brawley's police commissioner consumed alcohol and sanctioned bootlegging.[53]

Like gambling, the sex trade was popular and caused consternation among white elites. A large migratory labor force and a preponderance of males lured white, black, and Mexican prostitutes to the desert. In one incident, police caught Carrie Smith, a white woman, with a Native American man. Smith begged for mercy, arguing she had children to support. Smith received a suspended sentence, but her client got thirty days on the chain gang and a fine. The economic boom of World War I brought an explosion in moral infractions. El Centro's mayor claimed the town had ten prostitution dens where liquor flowed unabated. Addressing the rise in crime, the county passed a red light abatement ordinance. In 1924, Brawley cracked down on the sex trade, jailing dozens of prostitutes, many from Los Angeles. Claiming "negress" Pearl Ocampo was the town's "Tenderloin ruler" and had "been the source of much trouble and annoyance," police arrested her for operating a house for "evil purposes," a rap she often beat. In 1929, scores of prostitutes were expelled from the county and thirty Mexican "underworld women" were deported for violating the antiprostitution stipulation in the immigration law.[54]

The valley's penchant for morality ran headlong into the vice industries across the border. Building on Governor Esteban Cantú's earlier policies of licensing drugs, alcohol, gambling, and prostitution, Governor Abelardo Rodríguez expanded tourism and gaming in the northern district of Baja California. As a result, many settlers faced financial ruin and scandal when they became addicted to the nightlife in Mexicali and Tijuana. In 1924, repeated petitions by "upstanding" citizens persuaded the US government to close the border at 9:00 p.m., thus initiating the beginning of the end of the rampant vice trade on the borderlands.[55]

The desert's rules of racial etiquette were in constant flux with changes in the farming economy, immigration restrictions, and the agency of ethnic and racial groups. During the first decade of settlement, race relations were fluid. The preponderance of adult males created a masculine unsettled world. When Richard Laughrin's family moved from Oklahoma in 1911, family members drove wagons and packed cantaloupes for Japanese. Laughrin remembered Japanese bought fifty-gallon kegs of beer for workers, and whites invited Japanese neighbors to dances. Takanori Nimura's father told him that race relations were cordial in the early years of settlement because there were few white women to police racial and cultural boundaries. For instance, his father and a prominent white farmer often drank together. "In the pioneer days," he remembered, "there were few women around. I think there was much better communication."[56]

The cotton boom in the 1910s ushered in new racial rules. The arrival of large numbers of white and black southerners upended the fluid state of race relations. White southerners demanded rigid racial separation, and segregation became the norm. In towns, the railroad tracks marked the color line, splitting separate residential and business districts. Racial segregation in the larger towns began in 1912. El Centro officials set aside a "negro quarter" on the east side of the tracks. Police harassment, threats of violence, and the difficulty in getting bank loans enforced spatial separation.[57] Two months later, in Brawley, a "race war" nearly exploded when whites rose "up in arms over an invasion" of Mexicans and African Americans who serendipitously bought lots on the town's west side.[58] Over the next two decades, ethnic Mexicans and blacks protested against segregation in restaurants, housing, and schools. In time, racial segregation became a strategy to divide working-class whites from people of color to minimize working-class alliances. Defiantly, people of color bought property on the white side of town just beyond the railroad tracks in interracial commercial zones. Movie theaters, stores, restaurants, hotels, and homes filled this racial middle terrain. By the mid-1920s, for example, twenty Mexican families in Brawley owned homes on the "American side" of town.[59]

Expanded cotton acreage and a growing dairy industry during World War I lured white and black southerners, Mexican migrants, and European immigrants. The new residents brought women and children, and soon racial groups clashed in the fields, in school, and on the street. A county official confided, "there is a good deal of prejudice. There are many [white] southerners here and they just seem to hate everybody of a different color." In spite of deep ethnic and racial antagonism, individuals formed bonds of friendship in school, in adult Americanization classes, at public athletic events, celebrations, and during commercial or criminal activity.[60] Year-round, thousands of single men roamed the valley traveling to the fields or looking for jobs. Police kept busy hunting down beggars, fraud artists, and "professional cripples." In one instance, thirteen blacks were jailed for "safekeeping" and then run out of the county. Racial mixing threatened the racially ordered society, and police broke up interracial socializing. In 1913, "A crowd of Japanese, with a few Mexicans and white men were arrested Sunday morning for lounging on the park."[61]

Judges used the vagrancy law to get rid of "undesirables" and force the unemployed to find work or leave the county. The vagrancy law gave judges discretion to punish unacceptable behavior, to jail, deport, or banish economic malcontents. A newspaper reported: "Under the vagrancy law" all the police had to prove was "that the offender had no visible means of support, or that he had the reputation of being a dangerous and suspicious character," to arrest him.[62] Indicative of the control authorities placed on "undesirables," Brawley's police commissioner promised to eradicate all gamblers, "questionable women," vagabonds, bootleggers, narcotics peddlers, and other criminals. He claimed the valley was "the dumping ground for these undesirables being run out of other sections, and we intend to put a stop to it." In another incident, the Calexico police chief announced: "I am going to inspect every queer looking stranger that lights in this town and if they cannot give an accurate account of themselves, they will be forced to walk the railroad tracks out of town at the point of a .45 caliber Colt."[63]

Community leaders and the county sheriff used the chain gang to punish criminals and field union organizers. "Since the organization of a chain gang in Brawley, the wave of larceny, burglary, and theft has subsided," proclaimed the town's judge in 1922. Courts adopted creative sentences to deal with vagrants and the unemployed. Petty offenders often received a "floating" sentence, an agreement to leave town on a suspended sentence. Four Mexican "loafers" had to labor for five days on the city streets for refusing field jobs. The judge suspended the sentence to give them time to find work or leave town.[64] Convict labor cleaned streets, dug canals, and worked on infrastructure projects.

Working-class people of color responded to cultural insensitivity and racial animus in a myriad of ways. In the early years of colonization, for example, a white man caught photographing two Indian men building an adobe house wrote on the back of the photo: "They did not like to have their pictures taken as in their ignorance they thought that pointing the camera toward them worked some evil influence over them. So one turned his back and I snapped the other as he turned to see what I was doing. And was he angry!"[65] In another incident, a young well-traveled Cucapá caught a white artist drawing his home and told him to stop: "You Americans and Mexicans have different ideas from us. You like that way of taking pictures and photographs, but we don't like it. It is contrary to our ideas to have pictures of our men, women, children, houses, or horses, because some day all will be dead, and the pictures remain and are bad memories."[66]

Exploited workers found ingenious ways to strike back at pervasive racism. In 1920, Brawley justice H. B. Griffin, a notorious Mexican hater, complained about endemic crop theft, and he charged that Mexicans were "almost exclusively the offenders." He remarked: "I have a twelve acre melon patch west of town, and I have never visited that patch late at night or early in the morning that I have not found people there, helping themselves." He estimated they pilfered the equivalent of a railroad car. "I do not object to my friends having a melon or so, but when everyone in the country goes to the patch in wagon, cars and buggies and carts the melons away, besides splitting them open, right and left, and leaving them to rot, I do object."[67] Griffin failed to recognize how workers' paltry wages compelled some to supplement their grocery list at his expense as a passive aggressive blow toward the misuse of his authority.

Frequently, people of color exploded in anger and resorted to violence when they were racially disrespected. In these cases, elites punished and banished the offender. "I'm Mamie Brown, from Na'th Carolina, and I'm not a niggah," was the announcement preceding a fusillade "of salt shakers and glasses, when the would-be colored patron was refused service" in a Calipatria café on the white side of the railroad tracks. Brown replied to the insult by tossing a sugar bowl at the proprietor's head. When the deputy marshal arrived he beat a hasty retreat when a tumbler cracked him in the head. After routing the employees she "stalked out of the place" to her home where she was later arrested. Upon hearing the charges, the "wild woman" demanded a jury trial, where she "kept up a rapid fire of remarks" exposing the moral hypocrisy of white elites who went to the "colored" side of town for prostitutes and liquor, "but us colored people ain[']t allowed to spend our money over heah" on this side of town. She was convicted, fined $200, and sentenced to sixty days in jail, suspended if she left the county by sundown. She took the deal.[68]

Two incidents reflect the ways women challenged racial insults. As Mrs. Georgia Gihn Singh sat in a car, a white man named Claude Campbell "made remarks reflecting upon her race which she resented," according to a newspaper column. Singh walked across the street and punched him. What transpired next was open to dispute. "Campbell denies that he struck back at the woman," wrote the reporter, "exercising himself only to the end offending her blows and keeping her at a distance." Campbell claimed her nose was broken when her face ran into his fist. When a white waitress at a drugstore refused to serve Tona Valoa, "a young white woman" married to a Filipino, Valoa complained to the proprietor who quieted her down by selling her a mug of root beer. Still seething, Valoa splashed the beverage into the waitress's face.[69] These wide-ranging examples offer a glimpse into how individuals responded to racial animus. These expressions of defiance showed how those marginalized challenged racism, discrimination, and middle-class Protestant Anglo conceptions of morality.

Colonizing the Landscape

Desert settlers and hired hands used a variety of machines to transform the landscape.[70] A "fresno," a six-foot land lever pulled by mules or horses, was the most important mechanism to carve the landscape. A fresno operation consisted of an operator walking alongside the grader, which was pulled by a team of two to eight mules. The driver held a huge handle attached to a curved blade. As the mules plodded along, the operator either lowered the blade to cut a high spot or lifted it to deposit dirt. Although only six years old when he emigrated from Mexico in 1924, Modesto Ibarra vividly recalls his youth and how his family altered the landscape. His uncles owned and leased two hundred acres outside Brawley. He recalls his uncles operating a hundred mules with fresnos leveling ground. In another case, workers for S. A. Gerrard Company, the largest grower-shipper in the county "conquered" fifty acres in the north end's "frost-proof belt" to grow winter tomatoes.[71]

When tractors knocked down sand hummocks and graded fields, the disturbed earth whipped up breathtaking dirt storms. Modesto Ibarra recalled dust clouds blotted out the sun. The dust storms blew so hard a driver could not see the animals pulling the wagon. Settlers depended on the instinct of mules and horses to navigate home. Richard Laughrin clearly recalled the region's ecological transformation. Richard's father, James, left Oklahoma in 1910 to scout out the county. The next year James relocated his wife and seven children on a ranch outside of Brawley. The Laughrins worked for Japanese cantaloupe farmers. Richard remembers a landscape of tiny farms, cowboys, cattle, mules, horses, dairies, and a sea of alfalfa. As a teenager, Richard's first job was leveling sand hills as tall as a house. He recalled the irrigation district gave away water to erode sand dunes and knock down rocky mounds.[72] In 1923, the IID announced rules for water use other than irrigation. For $1 farmers received unlimited water to leach alkali, fill in low spots or arroyos, prepare soil for seed, and to level sand dunes and high ground. A few years later, a newspaper article described how farmers shaped the desert. C. S. Allen hired five fresno teams to dig ditches between 320 acres of sand hummocks. Then Allen ordered water to erode the sand dunes. Flood leveling had the added benefit of leaching alkali from the soil.[73]

The Imperial Valley's competition with farmers across the international line propelled water and land policies. As farm acreage increased south of the border, settlers in the Imperial Valley (who used the vast majority of water in the Alamo Canal) faced the possibility of water shortages. The Imperial Irrigation District (IID) tracked water orders in the Mexicali Valley for testimony on All-American Canal legislation and to build up a water record for treaty negotiations to determine each nation's share of the river. In 1905, Mexicali landowners ordered water for 3,591 acres. Five years later, farmers irrigated 15,000 acres. In 1921, farms encompassed a whopping 120,000 acres. Acreage reached a high of 217,000 acres in 1925 before dropping to 156,000 acres two years later. Alarmed by rising water orders in Mexicali, the IID encouraged US farmers to maximize water orders. Settlers also increased water use by pressuring the government to expand homesteading. Excess water drained into the Salton Sea and flooded parts of the Torres-Martínez Indian reservation, yet the IID rejected Indian complaints and refused to compensate for property losses.[74]

Initially, farmers were unaware of invisible passengers in the water. The incredible amount of salt in the river became a cause of concern in the 1910s when farmers and agricultural specialists took notice of stunted crops and fields covered with alkali. They soon realized clay in the soil hampered drainage. At the time, few knew proper drainage leached the soil of excess salt. The scientific engineering of the environment went into overdrive. In 1921, the IID hired a drainage engineer who laid out grids every mile. Observation wells dug at regular intervals determined the underground water table, soil consistency, and density. The IID's drainage "war room" contained maps indicating where groundwater required drainage. Mechanical scoop shovels dug a valley-wide system of main drains feeding into the New and Alamo Rivers. Discarded water flowed into the Salton Sea, which the Department of the Interior had designated an agricultural sump (submerging part of the Torres-Martínez reservation). By 1929, the IID had 234 miles of deep drains for the valley's 550,000 farm acres. By 1940, the district had a whopping 1,200 miles of drains.[75]

In the 1910s, agribusiness interests turned to the state for assistance on a range of issues. This dependence on state and federal technical assistance was indispensable for the growth of truck farming and later for procuring a mobile labor force of American and foreign workers. Generally speaking, settlers jealously guarded local control, fearing government regulations and strict enforcement of state labor laws requiring prompt payment of wages, workman's compensation, and housing. Although agribusiness decried meddlesome bureaucrats, it aggressively sought governmental help to protect its interests, arguing that the state and federal government owed them for domesticating the desert. Farm leaders requested crop, pesticide, and insecticide research and appealed for drainage and marketing assistance. To research scientific knowledge of desert farming, the California Development Company sponsored a farm experimental station. In addition, the stockholders of the C & M Ranch in Calexico and the Colorado River Land Company in Mexicali employed agricultural advisors and enlisted the help of the Department of Agriculture. Additionally, in 1912, the College of Agriculture at the University of California, Berkeley opened the state's first scientific agricultural research station near Calexico. The county horticultural commissioner and the Farm Bureau, coordinated with state and federal agents, offered advice to improve crops and livestock production and to control insects and crop diseases. The federal government made no distinction between American farmers on either side of the line.[76]

Throughout the 1920s and 1930s, the horticultural commissioner and the field experiment station urged farmers to flood fields to leach alkali. To drain excessive salt, irrigation experts recommended installing underground tile to remove stagnant water and salt. Across the county, workers laid six-inch pipes six to eight feet below fields. When the All-American Canal went into full operation in 1942, it carried some 3.5 million acre-feet every year. The amount of salt draining into the Salton Sea increased exponentially. In 1964, for example, 3.3 million tons of salt passed through the aqueduct into the valley.[77] Local, state, and federal engineers cooperated to eliminate alkali from waterlogged fields. After ten years of working on the valley's drainage system, a professor of soil technology wrote a progress report. In 1923, residents approved $2.5 million in bonds for drainage. Within two years dredgers had dug

"Ruth" dredger removing silt from canal. Courtesy P1974_001_0010, Imperial County Historical Society, Pioneers Museum, Imperial, California.

112 miles of deep drains into the most waterlogged and alkali-afflicted areas. As a result, the water table dropped from eight to ten feet.[78]

In addition to salt, an estimated 180 million tons of reddish-brown silt coursed through ditches every year. Machines scooped out half the silt and piled it alongside ditch banks, creating towering dirt walls.[79] It cost $1.5 million annually to clean mud and debris in ditches and fields. By 1930, the IID operated eighty-seven miles of main canals and almost two thousand miles of laterals. Crisscrossing the desert floor, the canals resembled a brown wire mesh of thick chocolate milk. When Imperial Dam, the All-American Canal's headgate on the Colorado River, was built in the late 1930s, it had the world's first "desilting" plant, where four settling basins each captured 50,000 tons of silt daily.[80]

Eradicating "Pests"

Mastery of the environment encompassed perfecting drainage, eliminating excessive silt, and especially, eradicating plant diseases, insects, and land and airborne "pests." Plentiful water, miles of alfalfa, and fruits and vegetables attracted armies of bugs, worms, aphids, birds, rabbits, muskrats, and gophers. Johnson and Bermuda grasses choked crops and did as much damage as animals. The threat "against dangerous importations" propelled the county to establish quarantines and hire livestock and bee inspectors. Federal research of synthetic insecticides and pesticides exploded during World War I, and the government urged farmers and ranchers to apply poison to increase production. In the early 1920s, the county supervisors passed an ordinance to fight "weed infestation" countywide. Horticultural commissioner B. A. Harrigan pledged to wipe out "Johnson grass, wild morning glory and other noxious weeds" by hiring professional poison applicators. In 1929, a Florida Mediterranean fruit fly plague led to a pest control conference in Los Angeles. B. A. Harrigan remarked: "Control is not an utter impossibility" as long as the state remains vigilant. If the fruit fly made it to California, "the pest will be kept localized if not eradicated."[81] Agricultural experts exhorted farmers to saturate crops for human consumption with poison. In 1926, for instance, the county farm advisor oversaw a demonstration by the state farm extension service and an entomologist from the Department of Agriculture to kill "insect pests" with various sprays. A decade later, the county embarked on a massive "weed-control and eradication program" to eliminate Johnson grass, morning glory, white horse nettle, Russian thistle, and camel's thorn. Farmers and laborers spread 13,545 gallons of arsenic acid, 4,000 gallons of sulfuric acid, 200 gallons of "atlacide," and 450 gallons of oil.[82]

Horticultural commissioner B. A. Harrigan also trained farmers how to eliminate worms. In one season, butterfly caterpillars ate $100,000 in alfalfa.[83] Battling a different kind of worm, Harrigan informed lettuce growers of the danger of the cutworm. He suggested care in spraying or dusting crops because the residue could harm consumers. In order to "control insect pests" like the cutworm, Harrigan suggested using a toxic "formula for poisoned bait" and applying it right before planting: "Paris green [poison] 1 pound, cheap molasses 2 quarts, lemons 1–2 dozen, water 4 gallons, wheat bran 25 pounds." Grind and mix the ingredients and spread it over five acres in the evening. Wiping out lettuce-eating worms required a deadly elixir of lead arsenate.[84] Airplane crop dusting with poison brought a new method to combat mildew. In 1926, an airplane crop duster pilot sprayed 600 pounds of sulfur on cantaloupes "to conquer mildew" that "attacked" cantaloupes and imperiled the valley's most lucrative crop. To win the war against mildew, the University of California, Berkeley, loaned the county an expert in plants, and farmers paid his salary.[85]

In 1928, the county's plant pathologist announced a virulent strain of powdery mildew had devastated crops for the previous three years. The old technique of spraying sulfur worked well with peas or grapes, but when sprayed on cantaloupe vines and related crops it burned the leaves and did more harm than good. He declared that fighting crop diseases and pests was one of the major tasks facing agribusiness. Farmers funded a project by state and federal experts to find a solution. Experts in chemistry, pathology, agronomy, ento-

Imperial County Farm Bureau member transporting bags of grasshopper poison. Courtesy P2003_210_001, Imperial County Historical Society, Pioneers Museum, Imperial, California.

mology, and botany collaborated to engineer a cantaloupe resistant to powdery mildew. After scouring the world they found a "Hindu melon" resistant to powdery mildew. Experts set to work perfecting a hybrid melon that grew in the desert, was tasty, and withstood nature's scourges.[86]

In the 1930s, county officials fought a desperate war against a grasshopper plague. In August 1933, horticultural commissioner B. A. Harrigan reported the annual battle against "hoppers" was intense. Earlier that summer, farmers blended a cocktail of fifty-four tons of arsenic and bran and sprayed it onto fields and ditch banks to kill grasshoppers. The Reconstruction Finance Corporation paid workers to spread poison. Four years later, "hoppers" continued to wreak havoc, compelling the federal government to ship poison bran mixed with sodium arsenate. During the infestation, the county put three hundred men receiving public aid to work on the "grasshopper eradication program." In 1939, the county spread ten tons of poisoned bran every day.[87] At the time, farmers did not wash produce before it was packed, nor were there ordinances protecting livestock from consuming feed tainted with poison.

Settlers also engaged in a systematic effort to eliminate snakes, mountain lions, rabbits, gophers, coyotes, and birds. Evelyn Zinn recalled her father irrigating during the day to avoid snakes at night. County employees laid out hundreds of poison baits to eliminate coyotes that ate poultry and cattle.[88]

Farmers focused special attention on birds, rabbits, and gophers. Under the heading "Black Birds a Great Menace," G. E. Wiest, chairman of the Blackbird Control Committee, sent a letter to his congressman begging for help. Wiest was angry because the Farm Bureau failed to convince the Department of Agriculture to combat the flying pests. Wiest complained, "Blackbirds" terrorized the county throughout the year, gorging themselves on crops. Wiest expressed the ideologies of controlling the natural world when he wrote, "some of our neighbors have spent hundreds of dollars in ammunition, time and labor in keeping the pests off the fields," and some "of our neighbors are not able to rent their lands for this purpose, and others are afraid to put in the crop." He pleaded for government help to find an "effective means of control." A few years later, an agricultural expert wrote a California senator: "During two seasons the most experienced bird control expert of the biological survey spent" many months in the desert "investigating the blackbird problem," yet "efforts to develop a systematic method for their destruction" had failed.[89]

Under the newspaper headline "Alladin Like Desert Change," Lewis Kirkpatrick, a farm manager, worked "wonders" on an 880 acre farm. Yet flocks of blackbirds numbering in the thousands feasted on the crops. "If left alone," said Kirkpatrick, "they would strip every grain of corn" on the ranch. The farm manager hired "a man whose duties consist simply in fighting these pests." The blackbird hunter placed a dozen discs on poles in the fields. "At dawn the hunter goes out with his gun and beats on the discs with a hammer. This causes the birds to rise from the corn where they have been feeding." Taking aim, "the hunter shoots into the flock, killing some and scaring the others away. He repeats this operation at every disc, keeping it up until about noon." In 1926, the county launched a campaign against blackbirds nested in tules along the New River and the Alamo River. To drive them out, the county horticulturalist asked farmers with tules bordering their land to burn them and spread poison. Stung by criticism from bird enthusiasts, the county horticulturalist denied killing other birds.[90]

The most popular pest extermination program targeted gophers. Gophers burrowed below the surface of a ditch and when the ditch was full, water flowed through the hole and destroyed roads or fields. In the mid-1920s the Imperial Irrigation District and the county announced 10-cent bounties for each gopher tail in "a campaign of destruction" to save crops and ditch banks. Payment was given for at least one hundred tails tied in bunches of ten. Within three weeks hunters brought in 40,000 tails. A local newspaper promoted "gopher control," trumpeting it was a "wonderful opportunity" for boys to make money during summer vacation. One young trapper made a whopping $236 for killing 2,360 gophers; his catch would have included seven hundred more tails, but someone stole them. A newspaper editorial called the campaign a success because gophers were the county's most destructive rodents. Buying tails was expensive, but "if this pest is brought under something that resembles control, the money will have been well spent."[91] On the day before Christmas 1925, an editorial celebrated the elimination of 120,000 gophers since July. The editor urged the bounty hunters on: "More power to Horticultural Commissioner B. A. Harrigan and his army of boy trappers." Gopher hunters broke the record for kills the following month when 76,168 tails were traded for a dime each. "The number of orphaned baby gophers, who will likely follow their parents to the grave, can only be guessed at," another newspaper editorial gushed. Four weeks later, trappers collected $6,000 for tails. By that time, the county had purchased two million tails. The successful program prompted the county to continue it indefinitely.[92]

A year later, hundreds of professional and amateur trappers filled the offices of the horticulture commissioner to claim their bounty for gopher tails on the last day of the season. In line and milling about were Mexican and white men and women as well as boys and girls. In all, 37,648 tails fetched 10 cents each. So many hunters showed up the county announced a two-day extension. "There are a number of old men who are professional gopher trappers and thus make their living in the valley. Some of them have averaged $6 and $7 a day, as they are experts in catching the mischievous rodents." They traveled in an ancient truck with the inside covered with hundreds of traps. "Their home is on wheels, they camp by the wayside, going from ranch to ranch, as they rid one place of the pests seeking better fields for their calling."[93] The slaughter continued into 1928, when the *Calexico Chronicle* gushed the county had bought thirty-four miles of gopher tails (three inches per tail) "to exterminate these pesky rodents." In addition, hunters collected bounties on the equivalent of thirty-seven and a half miles of muskrat tails.[94]

Rabbits joined birds, coyotes, badgers, and gophers on the animal elimination list. Horticultural commissioner B. A. Harrigan's primer on "How to Destroy Jack Rabbits" offered a variety of rabbit poison recipes, poured onto vegetables or fruit or grain. Harrigan also noted many hunters swore that hot rabbit blood smeared on the trees served somewhat as a repellant. Poisoning simply could not kill rabbits fast enough, though. The effort to eliminate rabbits inspired organized rabbit drives, billed as a "community rabbit hunt," for sport. A newspaper outlined the rabbit hunt for interested sportsmen. Participants planned to leave Brawley and motor to a ranch northwest of Westmorland. Once participants took their places, "rabbits [would] be driven toward a V-shaped enclosure[,] which [would] concentrate them at the apex, where the guns and clubs [would] be brought into use." All year, hunters concentrated on wiping out rabbits since they ate cantaloupe vines. Officials warned residents not to eat rabbits for they may have eaten poison.[95] In his study of irrigated Idaho, Mark Fiege encapsulated the ideological implications of conquering nature: "Conquest constituted the essential first step that irrigators took in their efforts to reclaim what they saw as a dreary, arid waste. In myth, men subjugated or even destroyed wild nature, then resurrected it—redeemed it—in a productive form. Conquest was a necessary, moral act, but to male Idahoans it was an emotional redemption, a regeneration of the masculine self." Like Idahoans, Imperial Valley settlers reveled in violent acts like clubbing rabbits to death, and poisoning, trapping, beating, and shooting gophers. Fiege notes that rabbit drives were community endeavors where men and boys chased rabbits into corrals and then pounded and shot them.[96]

Horticultural commissioner Harrigan edited a booster pamphlet that discussed the economic and social aspects of the county. The authors revealed the ideological conception of control and cultural supremacy structuring society. "In the brief span of twenty years [the county] has been changed from a harsh, uninviting desert to a veritable Garden of Eden. The very trails over which the Spanish padres once trudged their way, are now hidden beneath mantles of green verdure—growing crops of every kind. Where scarcely a white man found habitation a score of years ago, now fifty thousand people, prosperous and happy."[97] In contrast to the rosy portrayal of desert colonization expounded by Harrigan's booster pamphlet, the war against nature had been waged at great expense to taxpayers and the environment.

Conclusion

Imperial-Mexicali Valley's history is the quest for power, control, and profits. Before money could be made, the Colorado River had to be dammed and diverted. Engineering the landscape came next. Profits, rather than rural agricultural democratic renewal, fueled the enterprise. Speculators raked in profits, while absentee landowners made nice returns from rented homesteads. The California Development Company's subsidiary Imperial Land Company promoted colonization, sold water stock, acquired and sold land, and marked out towns. Along the way the CDC circumvented the authority of the US and Mexican governments. The CDC's water and land policies established a template for power that merged seamlessly into a crusade to control or eradicate nature's pests to maximize profit.

Absentee and large-scale landownership and labor policies had a profound impact on the history of the Imperial Valley. Initially, colonization leaders anticipated establishing a white landowning and water-controlling elite, utilizing family labor supplemented with migrant workers. The rapid shift to absentee landownership and huge farms created a demand for plentiful, cheap labor, which caused a demographic shift. In the 1910s, farmers recruited a multiracial and multiethnic workforce and tenants made up of blacks and European, Asian, and Mexican immigrants. Proponents of agribusiness often used the latest scientific methods to squeeze profits from the earth. Achieving this objective depended on financial and technical assistance from local, state, and federal agricultural experts. Government farm agents conducted research and gave field demonstrations. Experts identified drainage and salt problems, developed mildew and "pest"-resistant crops, and provided advice and money to combat land and airborne insects, birds, and rodents.

Farmers imprinted verdant farms onto an inhospitable desert. The desire to subjugate nature involved leveling land, distributing water, draining water-soaked and alkali land, and eliminating silt and weeds. Next, settlers moved to eliminate bugs, worms, butterflies, aphids, snakes, birds, coyotes, rabbits, and gophers. A local newspaper editorial summarized the common fear of nature and the desire to dominate it. "The desert is conquered now, that terrible monster that in times gone by sucked the breath of life from many prospectors who perished on the sands of thirst, that tyrant that killed by slow torture the many who ventured to walk across the burning waste."[98] A local writer penned an homage to the colonizers for greening the desert:

> I am an empire reclaimed, for I have transformed
> the barren wastes into blooming gardens—
> arid lands have I fructified with living waters
> and made them to blossom in superabundance.
> I spell service to humanity, opportunity to the multitudes,
> abundance to the diligent.
> I have strong men and true, conquerors of the desert
> and captains worthy of any industry.
> I am Imperial Valley![99]

Taming, shaping, and manipulating land and water were only the starting point in a holistic pursuit of conquest. A philosophy of control and authority acted like a rivet, binding the entire social structure. In the next chapter, this theme will be extended in an examination of industrial irrigated farming.

CHAPTER 3

Racial Agribusiness

> But, my friends, this great and productive valley must be kept for the generations of white people who will follow us.
>
> Senator James Phelan, Brawley, 1920

The history of Imperial Valley agriculture is a story of consolidation of power and control in pursuit of profits. This chapter examines the transformations in desert-irrigated farming, in particular the popularity of labor-intensive crops and the concentration of farm acreage, absentee landownership, tenancy, and the rise of powerful shipping houses. The Imperial Valley's rise as a farming dynamo in the first four decades of the twentieth century mirrored California's ascendency to the top rung of agriculturally productive states.[1] In the 1910s farm profits skyrocketed with extensive crops, such as machine-harvested grains and hay, and labor-intensive cotton and truck crops (fruits and vegetables). In the 1920s, shipping companies rushed perishable crops to market. Crop and livestock organizations met annually to discuss distribution, marketing, prices, labor, wages, and pest elimination. Organization and consolidation led to vertical integration for maximum efficiency. The concentration of acreage and shipping in the hands of a select group of resident and absentee owners, banks, and corporations significantly impacted the economic, social, and political history of the county. American landowners in the Mexicali Valley, primarily the Colorado River Land Company, cooperated with farmers across the line to develop the cotton industry.

Racial competition and racial politics overshadowed commercial farming, profoundly shaping the development of racialized, hierarchical agriculture.[2] State and federal laws buttressed unequal power relations and maintained white supremacy. Desert landowners recruited Japanese, South Asians, and Filipinos as tenants. Asian tenants played an instrumental role in the development and prosperity of cotton and truck farming in the Imperial-Mexicali Valley. A cotton boom on both sides of the border in the 1910s spurred Imperial Valley farmers to recruit southern black and white tenants. South of the border, the Colorado River Land Company recruited Chinese from Mexico and China to farm and pick cotton. Thus, by the second decade of the twentieth century, American landowners in the Imperial-Mexicali Valley created a racialized irrigated desert farming borderland.

US nativists ended Chinese labor transmigration in 1882 and banned Japanese immigrant workers in 1907. In 1913, the California legislature passed an Alien Land Law, which outlawed farmland ownership by "aliens ineligible for citizenship," but permitted three-year farm leases. In 1917, Congress further restricted immigration from Asia, except for Japan and Korea. In 1920, California voters approved an initiative forbidding Asians from leasing farmland and owning stock in farm corporations. Within a few years the Supreme Court ruled against Asian naturalization on the grounds that the Japanese (1922) and South Asians (1923) were not racially white. In 1924, Congress set a 2 percent immigration quota per nation, banned all Asian immigration, and created the Border Patrol. Anti-Asian sentiment crested during the Great Depression when Congress paid Filipinos to return home.[3]

As profit-minded landowners in the Imperial-Mexicali Valley experimented with different racial groups to find the perfect tenant, state and federal assistance proved indispensable for the growth of desert farming. The government paid for technical assistance, marketing, border quarantine stations, and an employment office. During the Great Depression, the federal government funded construction projects, "pest" poison, and the destruction or reduction of crops. Additionally, the government built the All-American Canal project.[4]

Experimenting with Desert Agribusiness

Between 1901 and the United States' entry into World War I in 1917, farmers grappled with the complex matrix of weather, soil, irrigation, crops, and marketing. Growers replicated California's orchard and extensive grain crops such as alfalfa, barley, wheat, Milo maize, dates, grapes, and citrus fruits. Orchard fruit, however, took years to cultivate, bankrupting many farmers. Alfalfa and other livestock feed flourished and hundreds of dairy farmers, mostly from Europe, opened shop. Farmers planted alfalfa and cereals (corn, oats, wheat, barley, and rye), and fed tens of thousands of mules, horses, and cows. Farmers shipped hay and grain to Southern California dairies and feedlots.[5]

Dairies were hard work, but they provided a crucial niche for immigrants to gain a foothold in the farming economy. Ichizo Sano and Tanizo Nimura, for example, entered farming as dairymen and were among the hundred or so Japanese dairymen. After taking an irrigator job, Sano grew melons with a partner for a few years. In 1909, he leased an eighty-acre dairy, becoming, as his family claimed, "the first Japanese cowboy in America." Tanizo Nimura arrived in 1906 and later leased a dairy near Calexico. By 1917, he made enough to travel to Japan to marry. When Nimura lost his business in the post–World War I recession, he and other Japanese leased land in Mexicali, where some continued as dairy farmers while others grew cotton. In 1929, Roy Westmorland coaxed Nimura and other Japanese to rent his large ranch. Nimura farmed Westmorland's land until he was interned during World War II.[6]

In 1910, a total of 1,322 farms existed in the valley. Although landowners held title to 223,602 acres, only 176,000 acres were actually cultivated, and they were valued at $20.5 million. Farms averaged 133 acres and were worth $88.70 an acre. The census counted 1,009 US-born white farmers, 135 white immigrant farmers, and 101 nonwhite farmers. Farmers fell into three categories: owners, tenants, and managers. Property owners operated 824 farms (62 percent) encompassing 136,926 acres. In all, there were 684 US-born white landowners, ninety white immigrant owners, and fifty nonwhite (Asian and black) farm owners. Tenants held 421 farms (32 percent) totaling 65,145 acres. There were 325 US-born white tenants, forty-five European immigrant lessees, and fifty-one nonwhite renters. Farm managers made up the third farm group. Seventy-seven managers presided over 21,531 acres (6 percent of land); sixty-eight of these managers were white Americans, eight were European immigrants, and one was a nonwhite manager.[7]

In the first decade of settlement, profits came mostly from cereals and hay. Perishable crops (truck crops) were a relatively minor component of the economy. Enterprising growers experimented with labor-intensive crops such as cotton, cantaloupes, watermelons, strawberries, and rice. Truck farming soon transformed the desert into a garden economy and diversified the planter class.[8] Desert farming wore down hardy farmers in a Darwinian contest of survival of the fittest. Growers faced a host of problems. In addition to heat, floods, droughts, and excess silt and alkali, farmers struggled with a lack of capital, land assessments, water stock payments, taxes, railroad rates, labor scarcity, and nature's "pests." Successful farmers and tenants mastered the five "Cs" to survive: capital, credit, contracts, contacts, and color.

Capital furnished equipment, supplies, living expenses, taxes, water, and labor. In the early years, farmers relied on local and Los Angeles banks for loans. As farming expanded during World War I, farmers lobbied the federal government for money. The federal farm bank in California, however, refused lending to the Imperial Valley because of the "flood menace" and the problem of "insufficient drainage" for crops. Representatives of the federal bank said loans would only be forthcoming with a dam project "controlling flood waters" from the Colorado River. Thus the valley's negative credit environment explains why settlers appealed to Congress for federal bonds to build the All-American Canal in 1919. Agricultural leaders believed with the backing of federal bonds to build the aqueduct, the Federal Land Bank would begin lending.[9] The inability to get federal loans profoundly shaped desert farming. Tenants had to rely on landowners and produce distributors and shippers for loans, resulting in a loss of control to outsiders who preferred large farms to small family farms to maximize profits.[10] Credit flowed to ethnic and racial groups deemed trustworthy. In a multiethnic and multiracial society where people often viewed each other with suspicion and hostility, once a group became stigmatized as untrustworthy, the entire group faced almost insurmountable obstacles to financial and social upward mobility. To discern why Asians, and to a lesser extent African Americans, were permitted to lease land—a step up from manual labor—and why Mexicans were relegated to field labor, it is important to delve into cultural and social stereotypes regarding reliability, honesty, and skill.

The opinions of three farm leaders illuminate the trajectories of power supporting certain racial and ethnic groups while punishing others. A researcher left a record of J. A. Harris, who owned the gins in Imperial County. He "secures money from Los Angeles which he loans to the cotton producers at 10 per cent interest, taking as security a mortgage of the crops. In this way he exercises a considerable amount of control over them." Harris made assessments based on racial ability. White farmers often went bankrupt, making them a poor investment, while South Asians, on the other hand, made profits.[11] In a tenant system, in which ground was leased for two or three years, turnover was high and creditors carefully screened applicants. Once a person of Harris's stature formed an opinion of a group's credit worthiness, his assessment held tremendous weight when landowners went looking for tenants. The sales manager of Earl Fruit, one of the oldest and largest shipping companies in Los Angeles, and an important company in the Imperial Valley, explained why landlords preferred Japanese. They were "the most industrious people" you'll ever meet. "They can make money in a place where a white man would starve to death." His company had farm "experts in the field all the time to advise" white tenants, "but we can't seem to get much results" out of them. Dr. E. E. Chandler, an absentee owner who leased ground to South Asians, believed they were honest, and whites trusted them. A merchant told Chandler that South Asians always paid their debts, but he lost money when Mexicans defaulted on loans. In a face-to-face society, credit was essential for survival. Once Mexicans developed a reputation as defaulters, credit for them dried up.[12]

Contracts were a third necessity for successful desert farming. Homestead applications were government contracts with stipulations that included citizenship or naturalization papers, which privileged white Americans. Nativists also prevented immigrants ineligible for citizenship from signing farm contracts. The 1913 Alien Land Law barred immigrant Asians from purchasing agricultural property, and the 1920 Alien Land Law outlawed farm lease contracts. After passage of the 1920 Alien Land Law, Asian immigrants entered into the unpredictable world of informal tenant arrangements. Farmers also signed contracts with shippers spelling out the terms of the sale, when the crop should be harvested, and loans. Farmers with bad reputations were at a disadvantage in negotiating a deal. Personal contacts were also critical to obtaining credit and leasing land. Absentee landowner E. E. Chandler lost money when his white tenant planted barley. When South Asian farmers approached him "in rather difficult circumstances," he leased ground to them and earned $9,000 in 1922 and $10,000 two years later. Chandler preferred South Asians because "the white man is not willing to do the hard work needed in the Imperial Valley. The Hindu will get up at one o'clock in the morning and haul his cotton a distance of thirteen miles to the gin with his four-horse team and be there ahead of the American who lives nearer to town. The latter does not like this."[13]

Color, synonymous with race, made up the final farm variable. Whiteness had numerous advantages, as well as disadvantages, for prospective tenants. Asians endured prejudice and discrimination, yet they pushed into farming with hard work, skill, white allies, ethnic crop associations, and the willingness to tolerate a substantial amount of racial enmity. White tenants had first dibs at leasing ground by virtue of their whiteness, but a dismal farm track record put them in disfavor among creditors. Farm leaders frequently replaced "poor white trash" with black and Asian tenants. Black southerners were experienced cotton farmers, while Asians had a sterling reputation. The Colorado River Land Company in Mexicali adhered to the same standards with regard to tenants. A 1923 US Department of Commerce report of Baja California stated that Chinese "are noted for their faithful compliance with their [leasing] contract terms" and "are considered first-class credit risks from the moral standpoint."[14]

Landowners, shippers, and bankers believed Mexicans were racially and culturally inferior, lazy, and untrustworthy, all qualities that disqualified them as agribusinessmen. Another significant factor working against Mexicans was that employers wanted to keep them locked into a farmworker caste. Thus Mexicans of modest means rarely had a chance to develop a reputation as successful capitalist farmers. The vice president of a local wholesale business commented: "The Mexicans are no good at all, they don't know anything about farming." Not only did Mexicans lack knowledge of mechanized agriculture, he believed, they were even incapable of manual labor: "You have to have some brains even for that."[15] Mexicans could not even get a start in the Mexicali Valley because American landowners feared giving them a farming foothold. The Colorado River Land Company justified white and Asian tenancy by dismissing Mexican initiative and intelligence, attributes that emerged in the 1923 US Department of Commerce report of Baja California, which noted Mexico's efforts to colonize its citizens "have been almost universally defeated by the lack of ambition and education and the general apathy and indifference on the part of the average individual affected." The report's section on Mexicali concluded with a patronizing statement: "This class of Mexicans, unfortunately, seems to lack the initiative, the spirit of energy and organization, and the persistent application necessary for such development under existing conditions."[16]

Desert Irrigated Farming

Irrigated desert agribusiness underwent a transformation in the 1910s with the expansion of crop acreage, an uptick in absentee landownership, widespread tenancy, reliance on farm managers, the experimentation with different ethnic and racial labor groups, and the beginning of large-scale migratory labor. Agribusiness interests responded to these changes by organizing and integrating. Shippers built dozens of packinghouses next to the railroad tracks.

At the same time, farm leaders formed crop associations to discuss issues related to growing, harvesting, packing, shipping, and marketing livestock and produce. The growth of labor-intensive crops in the 1910s is evident in the explosion of cotton and cantaloupe acreage. When cotton farmers earned significant profits on 6,538 acres in 1910, the cotton acreage grew steadily, even enticing film producers to include black pickers in a movie about the plantation South. One ranch had fifteen miles of cotton equipped with its own labor force and tent camp. Growers dreamed of replicating the Colorado River Land Company's (CRLC) cotton kingdom in Mexicali.[17]

The CRLC's landholding stretched forty miles below the border, encompassing 832,337 acres, although after subtracting the mountainous section, a total of 668,000 was farmable. Flooding and vegetation, however, rendered large areas too expensive to cultivate. After the floods, the CRLC transitioned from cattle ranching to leasing land for farming. In 1905, the first year of the floods, the company imported Chinese from the west coast of Baja California. Initially, the CRLC struggled to maintain a captive, inexpensive, foreign labor force. A stockholder wrote: "It is quite a hard matter to get the Chinaman to stay south of the line at small wages we are willing to pay, when the imaginary line on the north is the only thing between them and the high American wages if they can successfully cross and keep out of the clutches of the American officers." The company thought a solution was to partner with Chinese lenders to finance cooperative farming: "But if they were working for some of their own people it might be different." A stockholder in 1908 pitched a novel idea: "I will send to China and bring over coolies and get that part of the business started" in the Mexicali Valley. Within a decade, Chinese, and to some degree Japanese, tenants dominated 80 percent of cotton production.[18] In 1921, thirty-two Chinese rancherías (1,314 men and 128 women) encom-

S. A. Gerrard Company cantaloupe packing shed with "American Beauty" crate labels. Courtesy H79–26_3B41–2, Imperial County Historical Society, Pioneers Museum, Imperial, California.

passed 82,000 acres. Farming cooperatively, members contributed money to demonstrate their commitment and only took advances for food and clothing. Landowners and San Francisco lenders financed the tenants. After the harvest, the group paid its debts and split the proceeds.[19]

Sinophobia and opposition to foreign landownership converged during the Sonoran dynasty of Alvaro Obregón (1920–24) and Plutarco Elías Calles (1924–28). Pressured by agraristas (agrarian activists), the government seized property from foreigners and redistributed it to the landless.[20] The Colorado River Land Company's favorable relationship with the government allowed it to mostly escape land confiscation. In order to evade a 1926 law permitting the redistribution of land to campesinos (rural folk), the CRLC petitioned the government for permission to subdivide its estate instead, selling property to Mexicans and others who had money and experience with capitalist agriculture. The Agricultural Department and President Calles (who favored small privately owned farms rather than communal farms) agreed, and governmental functionaries suggested selling land to Mexicans or immigrants from Portugal, Spain, Italy, and France.[21] Drawing on its experience evading governmental dictates, the CRLC sold very few farms to Mexicans, although it complied with a law mandating 50 percent (later 80 percent) of all jobs be set aside for Mexicans. As a result, the company employed 8,000 Mexicans in 1928. Over the years, workers in the Mexicali Valley dug 3,205 miles of canals and added one hundred miles of ditches a year for the valley's 200,000 acres of farms. Illustrating the enormity of the cotton industry, the company stored

25,000 tons of cottonseed. In addition to trains stuffed with cotton bales, each year cotton mills processed one million gallons of seed oil and cotton by-products.[22]

When cotton prices spiked during World War I, desert farmers north of the border planted miles of the white fiber. By 1920, cotton fields exploded to 126,000 acres. Before the war, the Japanese government recognized California's expanding agricultural market and sent agents to buy crops. Japan soon purchased most of the cotton grown in the Imperial-Mexicali Valley. Landowners scrambled to recruit white and black Texas tenants and pickers. According to one old-timer, black "workers in some of the larger families pooled their wages and managed to buy small parcels of land and break into ranching and farming."[23] Imperial Valley farmers shifted to vegetable production when the cotton market collapsed in 1920.

Cotton's success led the emergence of profitable, speculative crops, like cantaloupes. By the mid-1920s, melons became the valley's premier crop. White farmers first planted cantaloupes in 1904. They grew wonderfully and farmers improved the methods of production. Yet white growers frequently went bust, primarily because of ineffective marketing. Landowners began leasing to Japanese in 1909, and within a few years they dominated the melon industry. Innovations the Japanese made included covering each cantaloupe with a white cap to prevent sunburn and maintain optimal temperature, and discovering ways to corner the market by ripening melons earlier in the year. The increase in cantaloupe acreage and the labor required to plant, cap, harvest, pack, and load the crates onto railroad cars was staggering. In 1915, workers packed a total of 8,156 acres of cantaloupes into 4,666 railroad cars. By 1920, this increased to 21,697 acres, filling 8,938 rail cars. Two years later, a whopping eleven million melons filled over twelve thousand railroad cars that if lined up would stretch 5.5 miles. By 1923, Imperial County grew more cantaloupes than any state in the nation.[24]

By 1920, the Imperial Valley was an agricultural powerhouse. Cantaloupes and other labor-intensive crops ballooned from 10,856 acres to 152,828 acres. As landowners increased the size of their ranches, labor-intensive agriculture compelled them to farm or rent their property in smaller units. In other words, landowners divided their property and rented to multiple tenants or hired farm managers. Census enumerators counted each individual farm unit—even if many farms had the same landowner—which explains the startling rise in farms from 1,322 in 1910 to 3,176 twenty years later. About half of the land was planted in extensive (machine harvested) crops. Between 1910 and 1920, extensive crops exploded from 76,276 acres to 207,281 acres. In 1920, the most popular were alfalfa (92,229 acres), barley (69,430 acres), Milo maize (35,590 acres), wheat (9,506 acres), and oats (526 acres). Acreage increased from 223,602 in 1910 to 347,467 ten years later. Newly planted homesteads pushed the land under the plow to 403,663 acres in 1930.[25] As the fame of the Imperial-Mexicali Valley grew, agriculturalists from around the world visited to study irrigated desert farming on a vast scale.[26]

When the 1920–21 recession decimated cotton, growers, with the financial support of shippers, switched to vegetables. As a report explained, shippers "assist the Japanese growers by advancing money for the growing of the crop, having a contract with the growers covering all particulars and it is some lengthy one at that." They lend "from $40.00 to $70.00 per acre[,] which means an expenditure of at least $400,000 by them to say nothing of the labor and expense by the growers." S. Kinoshita, a strawberry grower, illustrated the shift to truck farming. When he netted $5,000 for his five-acre crop, the local newspaper called him a fairy tale story. Kinoshita grew Klondike strawberries because they shipped farther without spoiling, were heavier, and had a brighter color and a sweeter taste. Although strawberries were one of the most expensive and hardest crops to grow, he handled the farm by himself.[27] Garden crops exhausted the soil within three years, so after two or three seasons of soil depleting crops, farmers planted "soil building" crops for three years to rejuvenate the earth. Japanese truck farmers moved out and white tenants moved in and grew "soil building" grain and alfalfa. Each passing year farmers became more proficient. They shortened the growing and harvesting season to reach niche markets and with a year-round growing environment, farmers planted two crops a year (double cropping) on the same field.[28]

Growers of Color

The importance of nonwhite farmers, especially Japanese and South Asians (called "Hindus"), to the success of Imperial Valley agribusiness is a little known but vital story. The catchall term "Hindu" (often used in a derogatory tone) stood for Hindus, Muslims, and Sikhs, religious groups from India. Most immigrants from India in the Imperial Valley were Sikhs and wore turbans, which inspired derision. Some had worked on British large-scale irrigated cotton enterprises where they learned the fundamentals of capitalism. Others were well-educated landowners or soldiers under the British. Punjabi culture prized patriarchy, individuality, competitiveness, hard work, and male social equality. Increased population, rising debt, and opponents of the British imperial system compelled them to settle in the United States and the Canadian Pacific Coast.[29] South Asians arrived in Imperial County in 1910 and helped harvest the first commercial cotton crop. South Asians moved quickly from manual laborers to tenant farmers. They compared the Colorado Desert to the Punjab, and by 1917 they held about seventy-five leases. They were aggressive farmers and gambled everything on a single crop. Their clannishness, highly competitive ways, religious cultural differences, and the desire to marry whites and Mexicans in order to acquire property sparked deep animosity and withering discrimination. Since few women from India traveled to the United States, whites alleged South Asians practiced homosexuality, and some were convicted of this crime. During World War I, hysteria over "alien" spies prompted one newspaper to comment: "It is also expected that the Hindu menace in the Imperial Valley will also be discussed by representatives of the County Council of Defense." South Asians tried to improve their public image by purchasing thousands of dollars of war bonds.[30]

Dr. E. E. Chandler, the absentee landlord, explained why South Asians were successful and why whites despised them: "The Hindu is quite serious and stoical, while the Japanese is more jovial and can take things as a joke in a better way." Japanese and South Asians were "business men and want to run their own establishments," said Chandler, and agribusiness leaders valued them. "Hindus could get money from the banks more readily than the whites—they work harder and spend less." South Asians also leased more productive land than whites, which bred jealously and hatred toward them. Japanese and South Asian farming success inflamed white competitors. A critic declared, "I am willing to pay 25 percent more for land next to my [white] brother, but would not want to live in the neighborhood of men of a hostile race."[31]

The California legislature passed the 1913 Alien Land Law to limit Japanese farming in California. It outlawed further farmland purchases and only allowed three-year leases. Asians responded by renting ranches and forming corporations to buy farms.[32] In 1918, for example, a newspaper predicted South Asian migration to the desert because whites surreptitiously sold them land. Newspaper articles focused on the economic challenges men from India posed, remarking they worked very hard and kept their costs down. South Asian success "enabled them to pay unusually high prices for land leases," which angered whites who were in competition with them to rent farms.[33] In 1919, the Imperial County Farm Bureau issued a report listing why South Asians were a social "menace" and "undesirable." The report found only twelve had married. In addition to veiled references to sexual competition with whites and homosexual activity, the report alleged the South Asians overburdened the courts with lawsuits. At a huge rally, white landowners signed a petition not to lease ground to Asians, prompting half of the South Asians to move to the Salt River in Arizona to grow and pick cotton. Contrary to the hysteria, white farmers vastly outnumbered Asians. In 1920, there were only 387 nonwhite growers, compared to 2,456 white farmers. Farmers of color held a paltry 27,618 acres in contrast to 319,804 acres farmed by whites.[34]

Asians were fantastic growers, and their success brought landowners profits, poured money into county coffers, and enriched merchants. In addition to their economic success, and contrary to negative stereotypes, Asians were model residents and were remarkably restrained in the face of discrimination and abuse. As persons ineligible for citizenship, living in a hostile and violent state, they made difficult decisions to remain on the land, including a willingness to tolerate racism to avoid racial hostility, deportation, or violence.[35] A report summarizing the ingenious ways owners and renters conspired to evade the law offers insight into racial tenancy and competition. Parties evaded laws through handshake deals, "excessive wages," gifts, fictitious corporations, and white directors. It noted, "Caucasian lawyers and others are acting as legal guardians until these racial Orientals of American birth become of age." The report concluded that Asian farming had created a situation where "in the opinion of thoughtful California pioneers" in a decade so much land "will pass into the hands of second generation Orientals" that they will dominate California agriculture. An absentee landlord dissented: "In reality, I do not believe the Imperial Valley is a white man's country and I am willing to hand it over to the Hindus and Japanese. The Negro is not nearly so industrious: when he has accumulated a small bank account he moves to town."[36]

In addition to reaping profits for landowners, many farm leaders tolerated Asians because they were not a social threat. They stayed on their ranches and rarely purchased town homes. They tried to blend into society by wearing western clothing and learning English. Most Asians were single males who rarely intermarried. Some Japanese married picture brides, while others traveled to Japan to marry. Some South Asians and Filipinos, however, married Mexicans, and in rare cases, white and black women. Many Japanese converted to Christianity, while South Asians married to Mexicans allowed their children to be raised as Catholics. Ray Chikahisa recalled his mother, Kayo, attending the Japanese Christian Church, sitting through services though she barely understood English.[37] Asians who bought farmland placed the title in the name of their wives, children, white friends, business partners, and lawyers. Doso Tokeshi and his wife, for example, requested permission from the county clerk to borrow $3,500 against land (worth $9,000) they purchased in 1929 in the names of their four children. With an attorney's help, the family secured a loan and purchased half a section and planted vegetables. Part of their appeal was that they planned to stay in America, evidenced by giving their children "American" names and sending them to "American" schools. Some immigrants from India, however, put property in their children's name to avoid losing it in a divorce.[38]

Newspaper articles shed light on women farmers of color like Marie Antoinette Giraudo and Teru Sanbonmatsu. In the late 1920s, Marie Antoinette Giraudo, the county's "cantaloupe queen," filed a complaint against nine Filipino cantaloupe workers for extortion, after they allegedly threatened her with clubs and guns if she did not pay them $800 in back wages. Three years later, during the Great Depression, Giraudo found herself in jail for thirty days for refusing to pay workers.[39] Another story about a female farmer of color ended on a more positive note. Teru Sanbonmatsu, a Japanese picture bride, married Yoshimitsu Sanbonmatsu, a Holtville grower. When Yoshimitsu died in 1931, in the depths of the Great Depression, Teru was left with five sons, the eldest only eleven years old. Undaunted, she filed with the superior court to administer her husband's estate, arguing she had assisted Yoshimitsu in running his large farming, packing, and shipping company. Teru petitioned to take over the business because she did not want to dishonor her husband's name since he died with $30,000 in debt, and, after all, she had a family to support. When Teru closed Yoshimitsu's financial affairs with the court five years later, she had transacted half a million dollars in business. She paid off half of her husband's debt by making the shipping transactions herself. An astounded reporter noted she accomplished so much "despite a disastrous period of unsatisfactory marketing conditions, depressed prices, loss of accounts, bad debts" and rising overhead.[40]

Filipinos also broke into farming in the early twentieth century. As American "nationals" they could travel to the United States, but could not become naturalized citizens. Like Chinese, Japanese, and South Asians, they started out as farmhands and worked their way into tenancy. Absentee landlords rented ground to Filipinos and financed them to grow labor-intensive truck crops. The rabidly anti-Asian Farm Bureau launched a movement to prohibit Filipino farming in 1936. The local chapter passed a resolution "discouraging the leasing of real property" to Filipinos. Responding to the resolution, the district attorney wrote the state attorney general inquiring if Filipinos could rent or own farms. The state's lead lawyer replied that Filipinos born in the Philippines of "white blood" and military veterans could own or rent farmland. Earlier, the legislature had defined Filipinos as members of the "Malayan race," thus forbidding them from marrying whites. In 1939, state attorney general Earl Warren declared leases to Filipinos violated the Alien Land Laws because they were ineligible for citizenship. Filipino tenancy increased despite these exclusionary efforts. In 1939, Filipinos organized a grower association in the Imperial Valley and celebrated its formation with a Fourth of July dinner.[41]

Of all the immigrant racial minorities, Mexicans were the least represented among growers, dairymen, and ranchers. There were five major reasons why landowners generally refused to lease or sell farmland to Mexicans. First, in Mexico, the vast majority of campesinos were subsistence farmers; they lacked experience with or knowledge of farm machinery, loans, crop associations, marketing, and shipping. Moreover, many migrants were miners or worked in urban jobs in Mexico and had no farming background. Leaders in Mexico recognized these deficiencies and they encouraged emigration as a strategy to teach campesinos capitalist scientific farming before returning home to

The Sanbonmatsu family, from left to right: Kozo, Teru, Mits, Yoshimitsu (holding Yoshiya), and Yoshiro, circa 1926. Courtesy of Bruce Sanbonmatsu.

rebuild the nation. Racial and cultural stereotypes in the United States also played a role. Agricultural leaders viewed Mexicans as "peons" lacking intelligence and initiative to operate a modern farm.[42]

Class considerations also factored in to minimize Mexican farming. Agricultural leaders strove to maintain a highly structured racial and class hierarchy by holding them in a caste-like state as farmworkers. By the 1920s, whites described Mexicans as "birds of passage" who enjoyed migratory life and never became citizens.[43] Additionally, the vast majority of migrants viewed their stay in America as a temporary or a seasonal sojourn to earn money and return home. After the Mexican Revolution, the Mexican government enacted policies intended to maintain the loyalty of its expatriates and encouraged repatriation. As circular transnational migrants, few Mexicans coveted United States citizenship.[44] Family decisions also influenced why few Mexicans became farmers. The devastation of the revolution motivated expatriates to send money home, thereby minimizing investment capital.[45]

In 1926, Paul Taylor, a professor of labor economics at the University of California, Berkeley, found only six ethnic Mexican-owned farms in the Imperial Valley. Two farmers were "middle class Mexicans" born in the United States, while two other farms were owned by working-class Mexicans (one was ten acres and the other two acres). About thirty Mexicans leased small plots of land and grew cotton and melons. When asked why more did not buy or lease farms, among the reasons given were: a lack of capital and the fear that attachments to the land would make returning to Mexico difficult. Others confided they could not afford to save money. Paul Taylor thought the preeminent reason was cultural: "Most of the Mexicans questioned as to why they do not buy" farmland like other immigrants "seemed never to have thought of it before, to regard it as an odd question on a matter of no particular significance, and gave not the slightest indication that the question set them to thinking." Paul Taylor concluded: "There is no indication that the rising generation of Mexicans born or raised in Imperial Valley is interested in either farm ownership or leasing."[46] Despite the desire to relegate Mexicans to field labor, the following examples reveal several successful Mexican farmers. In 1917, three wealthy Mexicans formed the California Farming Company with $50,000 in stock. They planned to grow alfalfa and sell it in New Mexico and Chihuahua, Mexico. Three years later a "well-known" Mexican family from Chihuahua, Mexico, operated a large ranch in the county. Joe Ramos, a successful Calexico grape grower, believed grapefruit and grapes would flourish in the desert. For ten years Ramos had managed some of the largest nurseries in the state. His experience working in a San Joaquin Valley cooperative convinced him cooperative marketing could be successful in Imperial County.[47]

In response to the plummeting demand for cotton, strawberries, and cantaloupes during the Great Depression, growers of color began diversifying.

South Asians, "the backbone of the cotton-growing industry," planted Milo maize, lettuce, and cantaloupes. Others busied themselves with dairying and growing alfalfa. Japanese farmers introduced new produce like tomatoes, celery, and eggplant in Niland's "Thermal Belt," an area known as being frost free. Japanese migrated in such large numbers to Niland that "new shacks" increased the settlement of "Little Tokio." By 1939, the interracial Japanese-dominated Niland Tomato Growers shipped eight hundred crates of celery and eggplant a month.[48] Other people of color weathered the Depression and continued operating in the farming industry. Sidney Johnson was a sixty-year-old "Negro cotton grower," and A. Gutiérrez, who made a small fortune raising sheep, purchased a large ranch and resided in one of the nicest homes in the north end of the county. Filipino farmers benefited from hard times when absentee landowners shifted from white tenants and turned to Filipinos to grow labor-intensive crops like tomatoes and asparagus. People of color also pushed into shipping, with four grower-shippers of color in Imperial County at the end of the Great Depression. For years whites tried to keep South Asians out of the profitable shipping sector, but by the end of the 1930s, South Asian customers enabled American-born A. G. Sanga to move into shipping. J. H. Grande, the largest independent lettuce shipper in the state, was another Mexican in agribusiness.[49]

Racial Tenancy

Agribusiness prospered, in large part, because of tenants, and understanding racial tenancy is critical to grasping the intersection of agribusiness, race, and power. Absentee landowners provided an opportunity for people of color to slip into the planter, dairyman, and rancher classes. Absentee landowners hired agents to recruit tenants and handle affairs such as paying taxes, further distancing landowners from land, tenants, and the community. Landowners and agricultural companies frequently offered loans to tenants. The half-dozen dairy creameries, for instance, advanced money to dairymen. Shippers financed tenants with capital and credit via chattel and crop liens. Likewise, landowners paid land taxes and water stock assessments in return for a quarter or a third of the profits.[50] In 1919, irate tenants formed an organization to represent their interests to landholders, with a spokesperson announcing: "Outside landowners are to be particularly interested in the organization, as it is claimed that half of the land in the valley is owned by men who have never been here to see their property. A yearly meeting is to be held and these men will be urged to attend." Deplorable housing conditions fueled anger. After a harvest, tenants chocked dirt roads moving farm machinery, horses, mules, and ten-by-ten shacks on wagons to the next leased farm. Some tenants owned multiple shacks, making it easier to move the kitchen, garage, and bedrooms. It is unclear what happened to the fledgling tenant association, but what is undisputable is landowners refused to tolerate any challenge to their authority. Racial animosity and racial economic competition were at the top of the list of reasons why tenants were relatively powerless against landowners. If American-born white tenants wailed too loudly, landlords could replace them with immigrants or people of color.[51]

A tenant union made sense. From 1910 to 1920, tenants increased from 421 (32 percent of farms) to 1,403 (49 percent of farms), farming 153,371 acres (44 percent of acreage). At no time between 1920 and 1940 did the tenancy rate drop below 39 percent. A 1922 investigation noted how "every agricultural commodity" was "more or less brought into existence by tenants." The report broke down tenancy by race and acreage: 2,200 Japanese farmed 28,181 acres; 495 men from India farmed 36,459 acres; 100 Chinese farmed 80 acres; and a few "poor white trash from Texas" farmed the rest, with blacks and Mexicans absent.[52] In 1924, commercial and farm representatives estimated 85 percent of all landlords lived along the coast or outside of California. A report noted four reasons for the astonishingly high degree of tenancy: "capitalists" were unwilling to farm themselves, land speculators were leasing land to nonwhites, absentee owners owned large holdings, and most importantly, the weather made it "impossible to successfully sub-divide and sell and operate in small tracts."[53]

White supremacists were in a quandary. Many grumbled that racial tenancy "as an institution is wholly bad socially in the Imperial Valley, for the class of tenants that is able to stand the excessive heat is undesirable and under no circumstances can ever assimilate with Americans. Japanese are more or less offensive, but the Hindus are positively and absolutely impossible." At the same time, however, the attempt to settle midwestern farmers was a failure, mainly because they could not take the heat. "Everyone knows that this will prove unsuccessful for in very few instances do substantial white Americans ever remain in the Imperial Valley during the entire harvesting season." Spokesmen for agribusiness meshed racial and environmental determinism to explain the failure of white farmers. Because of the heat and humidity, "about the 10th of June a general exodus occurs and the people who represent any financial responsibility leave and do not return until the 25th of September or the first of October."[54] In the racialized world of the white agribusiness elite, every racial group could till the desert soil and harvest crops except the white farmer and his family.

1920 Alien Land Law

California voters passed the Alien Land Law in 1920 to halt the gains made by Asians in the farming districts. The law outlawed Asian tenancy and Asian-controlled agricultural companies who purchased farmland. It was the state's second effort to reverse Asian gains in its most profitable industry. The law would have eliminated the desert's most successful tenants and replaced them with black or white farmers. Landowners and shippers fought the law as they had battled the Colorado River and nature's "pests." White benefactors joined with Asians to challenge the law in court and undermined it with handshake agreements. Seven years earlier, after the enactment of the first Alien Land Law (which prohibited future farmland purchases and restricted leases to three-year terms), many Japanese left the valley and migrated as far as Florida and Baja California, Mexico.[55]

After World War I, nativists, led by the Farm Bureau and the American Federation of Labor, rushed to break the grip held by Asian farmers in California. In January 1920, the California Oriental Exclusion League lobbied to close the loophole allowing Asian corporations to purchase farmland. Others called for a total ban on Asian farming. Under a newspaper headline in March, "Tri Color Map Shows Asiatics," a column noted: "A red, white and blue map of Imperial county, red showing land owned or leased by Japanese, blue showing Hindu holdings and white" for Americans was shown to county supervisors to pass a resolution in support of ending Asian farming. The tricolored map appropriately symbolized color and race ideologies in every aspect of life in the valley. Exclusionists circulated a petition to exclude Asian farmers.[56]

Myron Witter, publisher of the *Brawley News*, ran an editorial titled "The Hegemony of the White Race," supporting the Asian farming exclusion. As an absentee farm landowner he knew every angle of the debate. Witter stated that many on both sides of the Atlantic believed Asians desired global political and economic supremacy. He used a finessed argument for a racial preemptive strike, wrapped in a Social Darwinian argument. "In all struggles Nature provides but one rule, and that is, the survival of the fittest, and there would be no genuine and real danger until the yellow races prove themselves more fit than the white man."[57] Two months before the 1920 election, California senator James D. Phelan, seeking reelection, stood on a platform on Brawley's Main Street urging voters to support the anti-Asian initiative. The Democratic senator urged his listeners to protect the valley for future generations of white citizens. "We must turn it over certainly in as good condition as we found it," Phelan remarked, "and that can only be done by resisting Japanese aggression." Phelan used a time-tested argument to exclude Asians, arguing "successful competition with them is impossible, because their standard of living is so much lower than ours."[58]

The Japanese Association in Brawley responded to the law by announcing a meeting at a local school to take their case directly to voters, but it had to cancel the meeting when the American Legion protested the use of the school auditorium. Rebuffed, M. Shigematsu, the secretary of the local Japanese Farmers' Association, submitted a letter to the *Brawley News* titled

"An Appeal to Souls of Liberty." After asking why Americans were not satisfied with barring Asians from owning farmland, Shigematsu appealed to the readers' patriotism: "We have no right to fight politically in this country, but have a little voice to appeal to your reason and morals." Evoking the central American creed, Shigematsu concluded, "When all the Japanese are forced out of the country, the cantaloupes may thrive as they do today, and a lettuce ranch may be as green as today and the American nation will prosper with great power[,] but the American history will not be so glorious as it is today with the words of Liberty." Voters overwhelmingly passed the initiative; in Imperial County, the tally was 3,962 to 1,743.[59] Even as whites pushed Asians out of farming, Japanese contributed to the industry. For instance, weeks after passage of the antialien initiative, shippers wondered how U. Kojima's lettuce ripened so quickly. Kojima disclosed he experimented with a new variety of seed. A week later, growers and shippers were amazed that Kato and Frujimato harvested peas in November, since peas had never been shipped before January.[60] Some landowners replaced Asians with whites. Before the law, L. C. Lee of the American Fruit Growers rented 1,800 acres to Japanese. By 1924, he leased exclusively to whites, who, he claimed, rivaled South Asian cotton farmers. The sales manager for Earl Fruit remarked, "Down in the Imperial Valley all the lettuce and cantaloupes used to be grown by Japanese who were financed by Americans," yet by 1925, "practically all the lettuce" was grown by whites using Mexican labor.[61]

Once the initiative passed, Asians had a handful of choices: become farm foremen or managers, leave farming, sublease, have white friends hold land title, or put the land in minor children's names. Most signed "cropping" and "labor" contracts for farming. They also challenged antialien laws in the courts, with mixed results. In December 1921, a federal appeals court upheld the ban on cash leases, but allowed sharecropping contracts. The Japanese association cheerfully announced the decision would "be a practical method" to continue farming. The *Brawley News* agreed: "The primary purpose of the anti-alien land leasing law has been smashed to smithereens."[62] In May 1922, the California Supreme Court upheld the 1920 Alien Land Law, but ruled Japanese parents could hold property for their minor American-born children because failing to so do would violate the child's constitutional rights. A few months later, Congress passed the Cable Act, which stripped US citizens (aimed at white women) of their citizenship for marrying an "alien ineligible for citizenship." In November, the Supreme Court forbade Japanese naturalization. Early the next year, the high court ruled "High Caste Hindus" were racially Caucasian, but not "free white persons." Naturalized South Asians lost their citizenship and were placed under the antialien laws.[63] After the state legislature closed the "cropping" loophole, the Supreme Court upheld the law in late 1923. Within weeks, many South Asians relocated to Texas and Arizona.

Meanwhile, Japanese and white supporters conspired to evade the law. Statewide, district attorneys lobbied state attorney general U. S. Webb for "elasticity" to avoid "violent disturbances" in agribusiness. The Imperial County district attorney asked for an extension because five hundred Asians had signed leases.[64]

"THOUSANDS OF RICH ACRES CALL AMERICANS" screamed a January 1924 newspaper heading announcing a massive Asian exodus. The Associated Chamber of Commerce and the American Legion agreed the situation was "drastic," but it would be "beneficial" for the county in the long run. White elites painted "roseate pictures" of an all-white desert Garden of Eden. Alvin Jack, a farm leader who supported Asian exclusion, recalled how Asians made a comeback after the 1913 Alien Land Law. This time around they had to "be licked again" with vigorous enforcement of the laws. The battle over racial supremacy in California agriculture lurched back and forth. Some Japanese suggested following South Asians to Texas, while others considered moving to Baja California. News quickly reached the valley that Texans resisted Japanese settlement, and Mexico announced strict enforcement of its national quota on Japanese immigrants. Desperate landowners recruited black cotton farmers from Oklahoma. African Americans with large families moved in; buoyed by housing and loans from "financial interests," they spent freely at local businesses.[65] The county supervisors rushed to recruit white farmers by contracting with Smith and Rogers, a colonization company, to relocate midwesterners. Over the next year the colonization company brought prospective farmers to tour the desert. After looking over the countryside and talking to landowners, midwesterners complained about high property values and the refusal of absentee landowners to sell farms. In the end, the effort to resettle midwesterners ended in failure.[66]

Demolishing simplistic notions of race and power, and suggesting how racial issues were ferociously contested, landowners and Asians continued to challenge the antialien law. In April 1924, Imperial Valley farmer K. Noji tested the law again from a different angle. The case drew a great deal of attention, and the Japanese Pacific Coast consul visited the county to lend support. Weeks later, George Carter and Hakam Singh also challenged the amendment forbidding "cropping" contracts. Superior court judge Marvin Conkling (a landlord with Japanese tenants) issued a ruling in favor of Noji.[67] In late 1924, the US Supreme Court rejected four challenges to the state's 1920 Alien Land Law. Only three times between 1926 and 1940 did the Imperial County district attorney try to convict violators of the law, with mixed results. Dr. D. S. Saund, a South Asian who moved to the valley in 1925, benefited from lax enforcement. Saund worked his way up the agricultural ladder to become a grower. He became a citizen in 1949 and in 1952 voters elected him judge in Westmorland, the first nonwhite public official in Imperial County. Four years later, Saund, a Democrat, was the first South Asian elected to Congress.[68]

Farm Factories

The 1920 Alien Land Law altered farming in the desert. Eliminating legal Asian tenancy had the unintended effect of driving out small tenant farmers, accelerating land consolidation, increasing farm managers, and precipitating the rise of grower-shipping houses and large landed estates run by superintendents. Resettling white midwesterners failed. Instead of increasing white farming operations, more ground passed into the hands of absentee owners or agribusiness corporations, who hired farm superintendents, Asian foremen, or serendipitously rented ground to Asians. The Alien Land Law also began the dependency on national and transnational migrant field laborers. The culmination of racial policies against Asians also reduced local commerce and the development of social capital, as the law began to depopulate the countryside. The shift from small farms to corporate behemoths also changed hiring practices as growers moved away from hiring their workers and instead relied on farm contractors to bring in migrant laborers. This shift dehumanized workers and treated field hands as widgets in the farm budget. The change in the labor recruitment system made it psychologically easier to deny decent housing and to repress labor activism. Finally, economic decisions transferred from locals to absentee landowners and shippers.

In the wake of the Alien Land Law's passage, Chester Allison, a prominent engineer and Calexico banker who farmed both sides of the Imperial-Mexicali Valley, traveled to Japan to recruit wealthy financiers to colonize 100,000 acres of the Colorado River Land Company's property. In December 1922, Allison returned with a delegation of manufacturers and bankers, who anticipated selling the cotton to Japan and China. This scheme would have placed Japanese farmers next to the border and increased Mexico's share of the river, and a flurry of protest arose throughout the West Coast. A San Francisco newspaper warned the plan would place "an active enemy at the doors of the country," and this posed "a military danger" to the United States. Senator Hiram Johnson and Congressman Phil Swing asked the State Department to lodge a protest, and soon the parties involved abandoned the plan. In July 1924, a newspaper revealed Japan's Committee of Immigration and Colonization was negotiating to colonize thousands of Japanese in Baja California. Japanese settlement heightened settlers' twin fears of competition with superior farmers and the expected reduction in water from the river. Three months later, rumors of Japanese moving to Mexicali elicited outrage. The Mexican government bowed to the protest and canceled the plan.[69]

A prescient editorial in the *Brawley News* five years after passage of the 1920 Alien Land Law revealed a local economic and racial revolution. "This is probably the last year in which the individual small farmer will figure as an important factor in the cantaloupe deal. Rigid enforcement of the anti-alien land law, and the expiration of contracts made previous to various court decisions are rapidly taking the Japanese growers out of the market," remarked the newspaper. "In some cases they will continue to be employed as workmen and foremen, especially by the larger companies." The era of small vegetable farms was over because of the difficulty of mastering the science of irrigated desert farming. The editorial speculated the "situation will naturally throw the vegetable industry into the hands of a limited number of large operators, farming not only their own land," but also swaths of leased ground. "Whether the change in the method of farming is a benefit or a detriment to the community" was unknown. It was clear, however, that a farming "revolution" was underway.[70]

Less than a year later, the same newspaper printed another jeremiad about the destruction of the small landholder. "With the elimination of the lease contract form of operations, the big distributors, as well as the smaller ones, were forced to change their methods. Instead of contracting with many individual farmers, the big firms have been compelled to farm themselves, accumulating large staffs," which eliminated "the separate financing of the individual farmer, white or Japanese." The alien land laws accelerated consolidation, argued the newspaper: "Elimination of the Japanese except as foremen has largely been effected, but this has brought in its wake an economic situation wholly unexpected except by a few far sighted" shippers. The editorial cited data from the 1924–25 lettuce crop showing sixty-two companies and individuals planting over 24,000 acres on large tracts of land. Even more troubling, the editor recognized the depopulation of the rural areas. "On this vast acreage there will be few, if any, white families living except the necessary number of foremen and their families, while the great army of laborers necessary to properly till this acreage will consist for the most part of Mexican peons, Filipinos, and other itinerant laborers, here today and gone tomorrow."[71]

The rise of manager-operated farms during the 1920s signaled the flowering of corporate farming. Agribusiness in the 1920s expanded with more farms and land under operation, and the profits rolled in with plentiful and inexpensive water and labor, free state scientific data, and the lowest taxes in the state. In 1920, ninety-seven out of 2,843 farms had managers, a mere 3.4 percent, which totaled 10.4 percent of all acreage. The number of farm managers skyrocketed five years later to 356 (11.2 percent of all farms); they were entrusted with 17.2 percent of irrigated land. By 1930, there were 613 managed farms, almost a fifth of all farms and a third of the farmland. Managed farms had an assessed value of $22 million, while owner-operated farms were worth a mere $300,000 more. Tenant farms were worth $19 million.[72]

Another seismic shift in agribusiness came with the concentration of power by grower-shippers. In the 1920s, shippers consolidated farming, picking, packing, and shipping operations. Shippers used contracted labor to harvest and pack crops in sheds where produce was iced, loaded on railcars, and shipped. In 1924, lettuce shippers took another step in conquering the vicissitudes of the market and perfecting product standards by contracting with farmers to grow crops under a produce label to build brand loyalty and ensure quality control. Revealing the consolidated power of shippers, in 1936 the county agricultural commission compiled a list of fifty-one large grower-shippers and the crop acreage they controlled. Grower-shippers handled 58,486 acres, or 83 percent of the vegetable and melon crops in the county. Their gross income accounted for about $9 million of the county's $22 million agricultural income.[73]

From 1927 to 1931, the Imperial Valley was one of a handful of counties reporting "Good" economic activity, according to the US Chamber of Commerce's *Nation's Business.* In January 1932, the valley slipped into economic depression. The bankruptcy of agribusiness giant S. A. Gerrard Company, which owned and leased 12,000 acres in Imperial County, illustrated the humbling of agribusiness. The Cincinnati-based, vertically integrated powerhouse had farm superintendents, crop inspectors, packing sheds, and shippers. It hired white and Japanese tenants, along with white, Mexican, and Indian workers, and housed them in tents. In 1931, eight months before his company went bankrupt, S. A. Gerrard visited the Imperial Valley for the first time in a decade. Growers and shippers fawned over the "dean" of produce. Gerrard bragged that his operation had increased 400 percent in the twenty years it had been in the county growing produce and hay. In the past season alone he had shipped 3,500 railroad cars. The collapse of the cantaloupe market the following year shattered the company's finances. Government subsidies and federal policies eliminating crop acreage benefited larger farm operations, helping companies like S. A. Gerrard survive bankruptcy.[74]

The economic calamity, coupled with crop losses from strikes, crippled agricorporations and wiped out independent farmers. Exposing the havoc the economic downturn wrought, in 1933, the Imperial Irrigation District owned almost 21,000 farmable acres and held 262,580 acres in delinquent tax certificates of hundreds of growers.[75] During the economic collapse, farms decreased from 3,176 to 2,696, and gross agricultural output fell from $21.5 million to $13.7 million. Shippers claimed $3.5 million annual deficits from 1930 to 1934. In 1934, farmers and shippers presented financial records showing wages devoured 40 to 50 percent of gross income, 12 to 22 percent went to packing, and rent and taxes ate up 15 to 17 percent. Other overhead included machinery, repairs, seed, fertilizer, hauling, water, and miscellaneous expenses. A farmer named Mrs. Standlee, who grew wheat on 2,000 acres, despaired: "Not enuf to live & not enuf to starve."[76] Corporate interests took advantage of economic hardship by consolidating power. Like an anaconda squeezing its prey, banks, insurance companies, and other investors bought defaulted farms and strengthened their domination over farming, packing, and shipping. In 1934, a grower estimated that grower-shipping entities controlled 90 percent of the county's acreage through financing, leasing, or farming. One agriculturalist commented there were only "large operators [who] lease, own and farm for themselves—this is a big scale industry—the chief difference from ordinary industry is that we deal in perishable products."[77]

Agribusiness Associations

Imperial County agriculturalists were among the strongest advocates of business integration in the state, centralizing control over land, crops, packing, shipping and labor. The proliferation of crop and livestock associations helps explain why California became the nation's most industrialized agricultural state. A diverse climate allowed truck crops to be grown year-round. Associations and large shippers built packing and distribution centers, set prices and quality standards, determined the amount of acreage planted, shared technical information, and battled "pests." Working through labor contractors, they set wage standards and working conditions, minimized competition for workers, and crushed strikes. A 1929 newspaper editorial conceded the rugged individualist farmer did not fit reality since farming "is rapidly going into an industrial basis."[78]

Industrialized agriculture led growers and shippers into bonds of cooperation, but cooperation also smacked of socialism. Members of cooperatives recognized crop exchanges were rooted in the socialist concept of group sharing and profiting, so farm spokesmen deflected charges of hypocrisy by claiming agribusiness upheld "red blooded Americanism" and defended capitalism against foreign "isms." Agribusiness leaders rarely publicly admitted socialistic tenets. Farm spokesmen resorted to linguistic gymnastics to explain the contradiction between individuality and collectivism. In a speech to a service club in 1933, J. Cameron, of the Hay Growers Association, admitted the similarity of cooperatives and socialism. After summarizing the history of cooperatives and marketing, Cameron finessed the socialist foundation of cooperative ventures. Cooperatives, he stated, had a special "philosophy of marketing. It is the communistic tendency of our commercial society—we do not like the word communistic, but there is no other word that describes the state of mind." He pointed out anticommunist California, ironically, led the nation in cooperatives.[79]

Italians, Swiss, Germans, South Asians, Filipinos, and Japanese formed cooperative farming associations in Imperial County. Pooling resources began in 1902, when a man named Dr. H. Fuller believed a dairy industry would flourish in the desert. After getting a concession from Imperial Land Company to supply the region with dairy products, Fuller drove cows from the coast and opened a creamery. In 1904, he formed the Imperial Dairyman's

Association, and six years later his creamery produced 25,000 pounds of butter a week, with Los Angeles his largest market.[80] Early associations lacked scientific knowledge and market information. Farmers organized institutes to discuss ideas and gather information on how to control insects and "pests." After farmers clamored for state technical and marketing assistance, the state established an agriculture experimental station and the county set up a Horticulture Department. A local Farm Bureau organized in 1915 and within four years it had thirteen branches and seven hundred members.[81]

A study by Yoshio Ichikawa, *Cooperative Movement among Japanese Farmers in California* (1925), described the inner workings of two Imperial Valley cooperatives. At the time, a dozen Japanese cooperatives statewide enabled growers to pool resources, end gouging by middlemen, combat the Alien Land Law, procure capital, and deal with powerful shippers and the railroad. Ichikawa noted the Imperial Valley Agricultural Association, formed in 1915, was "looked upon as one of the successful cooperative undertakings among Japanese farmers in California."[82] By 1918, 250 Japanese belonged to the organization. The co-op published and distributed daily market conditions and prices. It sent inspectors into the fields and sheds and gave classes in efficient and proper picking, handling, and packing. It represented the association at the daily distributor's meetings, oversaw shipping, and forced distributors to publish their rules and regulations. After the 1920–21 recession Japanese experimented with new specialty crops like strawberries. After marketers defrauded them of $2 million, the Japanese permitted the co-op to market their crops. By 1925, it dealt with eighty-two distributors and a dozen market representatives, eliminating "many confusions and evils" in the old system.[83]

Across the border in the Mexicali Valley, the Colorado River Land Company, managed by H. H. Clark, established large-scale cotton production and a cotton cooperative. The CRLC recruited ginning companies to open shop in Calexico and Mexicali. H. H. Clark also organized the Mexicali Chamber of Commerce to support the cotton industry. When Imperial Valley farmers failed to copy the CRLC's vertical organization, Clark rescued American farmers by pushing specific types of cotton and urged farmers to form cooperatives. With an eye to growing the industry into a regional powerhouse, in late 1919 and early 1920, Harry Chandler, a major CRLC stockholder, held meetings in Calexico for Arizona and Imperial Valley cotton leaders. These gatherings bore fruit with the creation of the California Cotton Exchange.[84] Cotton exchange advocates petitioned the US and Mexican governments to permit the importation of Mexican pickers, and it standardized wages and marketed cotton. The cotton association served as a model for Imperial Valley's tomato, alfalfa and grain, dairy, citrus fruit, cotton, asparagus, cantaloupe, sugar beet, beekeeper, cattlemen, sheep, and hog associations. H. H. Clark expanded the Colorado River Land Company's borderlands profile when he became the president of the Calexico Chamber of Commerce in 1923. From this commercial perch, Clark urged Americans to emulate the Mexicali Valley's agri-corporate integrated structure. In 1924, Clark began promoting Acala cotton throughout Southern California, and he called for industry-wide cooperation between growers, lenders, ginners, buyers, and shippers. Then Clark helped create California's first Cotton Growers Association, and he became its first president. The cotton association reelected him in 1925.[85]

The evolution of agribusiness took a quantum leap when the American Fruit Growers and the Lettuce Growers' Protective Committee merged into the Western Growers and Shippers Protective Association (Western Growers) in 1925. Headquartered in Los Angeles, it spoke for Southern California agribusiness interests. Its first major triumph was lobbying the government to keep Mexico off of the immigration quota list in 1928 and 1930. Led by its executive secretary, Chester Moore, a longtime Calexico banker and the former manager of a migrant labor program, the Western Growers lobbied for Mexican emigration, marketed produce, set wages, and crushed farm strikes.[86]

By the end of the decade, unfettered farm capitalism saturated the market with crops and drove down prices to the point where almost everyone lost money. Unscrupulous farmers and shippers were mostly to blame because they refused to cooperate with the local office of the federal Bureau of Agriculture Economics to limit overshipping to certain cities. Cooperative-minded agribusinessmen argued for "pooling" shipments into a clearinghouse run by volunteers to discipline the market without the headaches of a formal association.[87] The movement to organize growers and shippers into cooperatives worked, and during the Great Depression they mushroomed in size and power. In May 1931, for example, a massive gathering of growers formed a clearinghouse "to control harvesting, packing, grading, inspecting and shipping of cantaloupes, honeydew and honeyball melons" as a way "to prevent a possible disaster to the industry." A committee had power "to control all orders for refrigerator cars" and "authority to eliminate shipments of undesirable sizes" of melons. Weeks later, with melon prices slumping, the clearinghouse made history by halting all shipments in the middle of the season to raise prices. When the market collapsed, the clearinghouse folded. State officials termed it "the worst marketing disaster Imperial valley ever experienced."[88]

In the midst of the largest strike in the history of the Imperial Valley, in 1934, the Western Growers linked with corporate and agribusiness statewide to form the Associated Farmers of California, an organization of bankers, shippers, large landowners, and insurance companies. Hugh Osborne, an Imperial County supervisor, banker, absentee landowner, and leader within the American Foreign Legion, came up with the idea of the Associated Farmers. After he established a local chapter, he traveled throughout the state organizing county units. As a result of Osborne's persistence, the Associated Farmers became arguably the most powerful agribusiness association in US history. It raised funds, disseminated propaganda, imported scabs, and managed thousands of vigilantes and spies to break strikes and agricultural unions.[89]

Farm Subsidies

Dependence on state and federal assistance proved indispensable for the growth of desert irrigated agribusiness. A psychology of entitlement permeated the minds of settlers, who insisted on government assistance to help conquer the desert. State and federal money and technical assistance poured into the county. The state provided funds and technical assistance managed by the agricultural college at the University of California, Berkeley, which developed crops, drained alkali land, and conducted pesticide and insecticide research. The state also furnished crop inspectors, "pest" and plant disease eradicators, an employment office, crop clearinghouse assistance, and quarantine stations at the state border. Federal patronage began with opening the public domain to homesteaders, a market bureau, favorable immigration policies to import plentiful Mexican transnational laborers, and special agents to defeat field strikes.[90] During the Great Depression, federal money built roads and bridges, dug and cleaned ditches, provided make-work jobs, built and staffed migrant labor camps, dispensed free herbicides and pesticides, and provided crop subsidies. The federal government topped off this list of largesse by building Hoover Dam and the All-American Canal.

Federal support on a significant scale began during World War I, when the Department of Agriculture opened a branch of the Bureau of Markets to market, sell, and regulate crop shipments through cooperatives. Over the next decade the government's marketing agency coordinated the lucrative cantaloupe and lettuce industries, bringing wealth and fame to the Imperial Valley. Yet farmers and shippers undermined the Market Bureau when they made their own shipping decisions. Shippers tried to control the market through cooperative pooling arrangements to distribute, market, and withhold surplus products to inflate prices. In 1922, however, a court ruled pooling monopolistic and therefore illegal. Dejected, shippers once again turned to the government for help.[91] Later that year, at the largest meeting of cantaloupe shippers in county history, distributors approved a resolution to ship only ripe melons. Farm leaders, representing forty-seven distributors and shippers, begged the Bureau of Markets to oversee the shipment of the record 31,000 acres of cantaloupes. Melon overproduction caused a market glut and prices below production costs. Farm leaders, under the direction of the Market Bureau, established procedures to bring order to farming and shipping.[92] Mired in cycles of debt and uncertainty, vegetable shippers loaned money to farmers who had no choice but to continue farming, even if prices were low. Even worse, growers harvested at the end of the season when the market was saturated. Saddled with debt, and with little or no capital to begin the next season, growers begged shippers for more credit. Farmers tried to get out of the cycle of debt and dependence on shippers by demanding the government take over marketing and distribution, especially limiting shipments when prices were low. Growers got their wish in January 1923, when the Bureau of Markets agreed to open an office in El Centro. A year later, the Agriculture Department reported the valley's cantaloupe harvest "the most spectacular movement of fresh fruit in the world." The Market Bureau moved on to manage successful lettuce shipments.[93]

In addition to market and distribution coordination, farmers received assistance from the state Bureau of Standardization, which sent inspectors to examine tomatoes, potatoes, cauliflower, cabbage, and lettuce to ensure proper packing of produce. Inspectors fined more growers of color for violating crop standards than whites, who only received warnings or suspended sentences. Supportive government bureaucrats also obliged the Imperial Valley's request to quarantine the state borders. In 1922, the state complied with a request from Imperial County for inspectors at the Yuma, Arizona–California border to check vehicles for plants or seeds infected with banned insects or fungi.[94] Four years later, farmers' perennial labor shortage came a step closer to resolution when the state Labor Commission opened an employment office in Brawley to employ transient laborers and alleviate labor shortages. Before the decision to open the employment office, workers gained wage bargaining leverage by refusing to work until crops started ripening and farmers had no choice but to increase wages. Farmers lobbied the state to remedy this labor tactic by opening an employment office. The local newspaper celebrated the good news, reporting now "all men idle on the streets would be compelled to register for work or leave the city. They would be sent out in turn to the shippers or ranchers that need them and if after two or three failures it was seen that they did not really desire employment they would be asked to leave [the county]."[95] The employment office undermined workers' ability to withhold labor as a bargaining chip to negotiate wages. The employment office is another example of an agribusiness subsidy that gave farmers an unequal level of power over workers.

Technical assistance became the centerpiece of agribusiness public assistance. For thirty years, state and county taxpayers supported the county agricultural station. When L. Gordon Goar, the superintendent of the experimental station, retired, agriculturalists heaped praise for his contribution to farming. In addition to investigating and promoting draining systems and soil quality, Goar and other experts offered a wealth of technical help. Throughout the 1920s he labored to conquer nature's "pests" and to develop perfect strains of fruits, vegetables, and staples. In the early 1930s, Goar's team developed two highly profitable crops: flaxseed and sugar beets. The county's flaxseed production broke world records in 1939 with 42,000 acres.[96]

Economic ruin during the Great Depression propelled farm leaders to insist on more government aid. In 1933, the Agriculture Adjustment Administration (AAA), under the auspices of the Agriculture Department, began the task of reviving the agricultural sector. In this case, reviving meant limiting crop production. The government aimed to reverse farm foreclosures by offering money to farmers in return for not planting crops. The "adjustment" program threw farmers a life raft in the form of cash "allotments." In addition to farm mortgage refinancing, the AAA boosted cotton prices by paying farmers to voluntarily restrict crop acreage (called fallowing). This peculiar form of government assistance bewildered farmers, and county officials struggled to explain the process of getting paid for not farming. Later, government programs paid farmers to destroy crops, which further undermined the ethos of the food producers. Farmers began viewing agriculture differently. Water, land, and food became objects in the capitalist system to profit from in a new way. Crops fed and clothed people, but when the government paid the farmer to limit or destroy their production, food lost its core essence as sustenance, and it began to be viewed as just another product made in a factory and discarded when prices were low. Cotton fallowing and subsidies were the latest iteration of the patron-client relationship between government and agribusiness.[97]

The AAA groped toward a policy to keep agribusiness afloat. Few desert farmers grew cotton, so this subsidy was ineffective. The government quickly set up a new program for wheat subsidies, which established rules and regulations for other crops. The AAA's effort to reduce wheat acreage showed how crop reduction and subsidies evolved from halting family farm foreclosures to enriching large landowners. For three years, however, the AAA had little impact on Imperial County because farmers grew little cotton or wheat. Agriculturalists finally hit pay dirt in 1936 when California lobbied for "soil building" and "soil conserving" crop subsidies. Government handouts were embarrassing since farmers were outspoken proponents of small government, personal responsibility, and low taxes. Farm leaders obfuscated cash subsidies with "soil building" and soil "conservation" euphemisms. The program gave the largest landowners the most money. Under a soil-conservation plan, growers received payments for planting crops such as alfalfa and shunning soil-depleting crops like wheat, barley, and vegetables. In July 1936, a total of 1,793 farmers completed conservation worksheets.[98] Imperial Valley's share of "soil building" crops totaled $323,964.50, the largest of any county in the state.[99]

By the end of the decade, corn, wheat, rice, and cotton were included in the "soil conservation" and "price adjustment" program. After tasting the nectar of direct financial aid, in 1937 agriculturalists angled for more public assistance by forming the Imperial County Agricultural Conservation Association, a lobbying group.[100] The Soil Conservation Committee met with the Department of Agriculture to review 1937 subsidy payments and subsidies for the next year. Ironically, the Department of Interior had just completed Hoover Dam and was in the midst of building the All-American Canal to increase farm acreage![101] Critics of the All-American Canal had warned against the Imperial Valley's strategy of World War I veteran resettlement on homesteads to garner votes for the project and to increase water use.[102] The excess land under cultivation resulted in crop surpluses, bankruptcies, water wasting (which flooded the Torres-Martínez reservation), and during the Great Depression, paying farmers to destroy crops or to reduce production. The IID's policies ruined the lives of many farmers and created an ecological disaster for neighboring Indians, while depriving Mexico of the water it needed to fully develop the Mexicali Valley and thereby provide jobs for its citizens so they would not have to live the hardscrabble life of itinerant farmworkers in the United States.

Conclusion

Irrigated desert farming proved extremely difficult. Among the litany of obstacles agribusiness faced were lack of credit; inexperience; and the high cost of planting, thinning, harvesting, packing, shipping, and marketing crops. Farmers also dealt with market uncertainty, early freezes, or blistering dry spells. Originally, settlers envisioned an agricultural economy populated with white family farms growing extensive crops and citrus. By the 1910s, an alleged inability of white farmers to withstand the heat, farm failures, and a desire to maximize profits led to increasing absentee ownership and tenancy. The abandonment of family farms dovetailed with the shift to labor-intensive fruits and vegetables, which increased absentee ownership and the tenancy system. The desire for profits trumped race and landowners rented to farmers of color. Asian tenants injected a racial dynamic into agriculture, and white supremacists passed the 1913 and 1920 Alien Land Laws to restrict Asian farming. Even though Asians were instrumental in establishing highly profitable labor-intensive vegetables in the desert, spokespersons for the farm elite, like Imperial County district attorney Elmer Heald, continued to deny their contribution: "Our white men are the best farmers in the valley."[103] The 1920 Alien Land Law had the opposite results from its intention when it pushed white tenants off the land and reduced, rather than enlarged, the number of family farms. Absentee owners and corporations purchased more land and hired farm managers (who often hired Asian managers or subleased to farmers of color). The new system altered the farmer-worker relationship. Before, small farmers and tenants worked the land and personally hired harvest labor. Now, farm managers or shippers hired labor contractors to recruit field crews, thereby distancing the farmer from the worker.

Throughout the 1920s and 1930s agribusiness vertically integrated with the help of crop and livestock associations and cooperatives. Associations and cooperatives pooled shipments, handled marketing, set quality standards, disseminated scientific data, recruited workers, and set wages. They enlisted the cooperation of federal, state, and county agricultural agencies. This concentration of power transformed farms into field factories: mechanization, scientific farming, and vast estates owned by absentee owners relying on farm managers or tenants employing thousands of local and migratory workers. Landed interests, shippers, banks, and insurance companies controlled water, land, lending, and transportation to hold the county by the nape. Author and historian Carey McWilliams summed up this concentration of power when he wrote that in the Imperial Valley, "the soil is really mined, not farmed."[104]

To protect and improve truck crops, farm leaders demanded government assistance. In order to maximize efficiency and profits, shippers built centralized packing and shipping houses to rush produce to market. Dependence on the government grew during the Great Depression. The New Deal's soil reduction program paid farmers to destroy crops or to fallow ground. This policy violated farming's ethos. Destroying or reducing acreage depressed the local economy and reduced county tax proceeds. Even more egregious, it had a staggering impact on starving laid-off workers, who saw wages disappear under the wheels of a tractor. Historian David Kennedy summarized this phenomenon best: "The New Deal laid the groundwork for a system of farm subsidies that in the end mocked the pieties of frontier individualism and made the agricultural sector a virtual ward of the state." He pointed out: "Save only swaddled infants in their mother's arms, no members of American society would emerge from the New Deal more tenderly coddled than farmers, especially those large-scale commercial growers to whom most New Deal agricultural benefits accrued."[105]

The final element of agribusiness success involved establishing the perfect labor system. Leaders of agribusiness determined that if they could control the Colorado River, engineer the landscape and crops, eradicate "pests," and maintain a hierarchical, racialized farming class, surely they could create an ideal workforce. The story of the establishment of a perfect labor force and its relationship to farming is the subject of the following chapter.

CHAPTER 4

Racial Labor

When there is no Mexican, there will be no Valley.
Agribusiness leader, 1926

In a glowing column titled "Mexican Labor Is High Class," the *Brawley News* could not contain its enthusiasm for the seasonal worker program agricultural leaders negotiated with the US and Mexican governments in 1920. "'It is a splendid class of labor which is being imported from the interior of Mexico,' is what those who have seen the Mexicans' work" are concluding. The field hands "have no education whatever, but they have come from the interior, where they have never had enough food to eat, enough clothes to wear, or a comfortable place to sleep, and for this reason it is said, are appreciative of the good treatment and desirable conditions they find here, and are therefore contented and good workers."[1] The newspaper's comments came after the creation of the first successful large-scale seasonal migrant worker program in the county's history. This newspaper column reflects the viewpoints of the growers and is part of a longer story of the formation of an agricultural labor structure in the desert.

This chapter examines the construction of an adult, masculine, racially segmented labor market in the Imperial-Mexicali Valley. Building on efforts to subjugate the Colorado River, transform the landscape, and establish profitable agribusiness, labor control was the fourth pillar in the foundation of desert irrigated farming. As the economy transitioned from machine-harvested crops to labor-intensive produce, farmers reformulated the workforce to maintain a year-round labor pool while employing labor contractors to recruit and manage crews of migrants. Close proximity to the international border, labor scarcity, race, nationality, class, and gender influenced the making of a racialized labor market. Racial politics marked and placed fieldworkers within the valley's social and economic pecking order.[2]

Agricultural labor history in this desert valley went through four phases, each lasting about a decade. During the colonization period, the mutual water companies and farmers hired white residents, itinerant whites, indigenous peoples from both sides of the border, and crews of South Asian, Mexican, Japanese, and European immigrants. The expansion of cotton and truck crops in the 1910s required a larger and different workforce. Farmers opened the second labor phase in the county's history by experimenting with Chinese, Korean, Filipino, black, and Puerto Rican migrants. In the Mexicali Valley, the Colorado River Land Company recruited Chinese from northwestern Mexico and China to farm and to pick cotton. The demand for food and cotton in the United States during World War I caused an even more acute worker scarcity. Agribusiness lobbied the US government for permission to import Mexican seasonal field hands.

After the war, Imperial Valley farmers moved aggressively to open a corridor into Mexico for transnational *braceros* (laborers). During this third phase in labor relations, race and gender played a major role in shaping the packing shed workforce. Japanese packing shed owners exclusively hired Japanese as the highest-paid agriculture workers. Japanese shed workers looked down on field laborers, and whites grew angry when they were locked out of shed packing jobs. Over time, white and Japanese packing shed owners hired white male packers and general shed workers. The 1920s witnessed the maturity of a segmented, tiered, race-gender-caste labor pool. Farmers hired resident Mexicans for year-round work, while labor contractors brought crews of Mexican, and occasionally Filipino, men during the harvest season. Meanwhile, during the 1920s, unemployed braceros in the Mexicali Valley protested against Asian workers, which resulted in the enforcement of a law requiring 50 percent (later raised to 80 percent) of all jobs be reserved for Mexicans. Imperial Valley's labor system reconstituted itself during the Great Depression when impoverished white and black American migrants and Filipinos arrived to complicate the county's race-class hierarchy.

Government agencies also played a role mediating relations between boss and worker. Low wages, inadequate housing, and disgusting living conditions contributed to malnutrition, starvation, and disease. In 1913, progressive reformers created the California Immigration and Housing Commission to facilitate immigrant adjustment and to investigate field labor camp sanitation and housing.[3] Although farmers welcomed government programs that benefited agriculture, they stridently opposed improving living conditions. State agencies were more successful in enforcing wage laws. Before the Great Depression, the state did little to improve the working and living conditions of fieldworkers. During the 1930s, however, the federal government attempted to recalibrate the employer-worker relationship by dispensing aid to the unemployed and mediating labor disputes.

Desert Fieldworkers

In the early years of settlement, the California Development Company and irrigation districts recruited indigenous inhabitants from Baja California, Sonora, Yuma, Arizona, and San Diego to dig ditches, level land, carve out town sites, and build adobe homes. Farmers depended primarily on Indian families to plant and harvest crops. During the floods, two thousand Indians closed the break in the riverbank. As acreage expanded, the scarcity of indigenous workers compelled employers to search for other sources of labor. By 1903, for example, white migrant men harvested cotton.[4] At the time, Japanese workers predominated in the state's truck crop and citrus industries, but farmers shied away from them because they had organized unions and "started the first phase of rural labor militancy, using strikes, job actions, slowdowns, boycotts, and other militant tactics" in the fields. By 1907, Japanese were the highest paid farm workers in the state. The specter of interracial unity was more troubling than the prospect of Japanese controlling the fields. In 1903, workers in Oxnard formed the Japanese-Mexican Labor Association to demand higher wages and labor unity during the sugar beet harvest. The union went on strike and overcame a murder and beatings to win wage increases and other demands. When the union sought a charter with the American Federation of Labor, its request was rejected on racist grounds, effectively dooming agricultural racial solidarity for three decades. Until World War I, agricultural leaders across the state struggled to end dependence on Japanese and to eviscerate racial working-class unity by procuring other immigrants of color and assigning different racial groups separate tasks at different wage rates.[5]

In 1909, Imperial Valley farmers desperate for field hands recruited grape harvesters from Fresno, and employers sent advertisements for black cotton pickers as far away as Louisiana. Although most farmers desired workers of color, labor shortages forced them to hire European immigrants, US-born white "harvest stiffs," hoboes, and town folk. By 1910, eight hundred farms paid $700,000 in wages. In the same year, growers advertised for pickers across the Southwest, luring blacks, Koreans, South Asians, Japanese, Filipinos, Puerto Ricans, Mexicans, and Indians from Arizona. Employers prized South Asians for their cotton-picking skills, and they were valued as an intermediary group to divide workers to prevent union organizing.[6] Across the line in Mexicali, in 1910, the Colorado River Land Company imported fifty-one laborers from China who landed in San Francisco and proceeded to Mexicali under bond. Determined to limit Mexican settlement and dependence on Mexican laborers, in 1913, the company sparked uproar when it fired 125 Mexicans and replaced them with Chinese, pushing the total from China to 1,600. The CRLC and other foreign landowners took advantage of the political turmoil during the Mexican Revolution to recruit about 7,000 Chinese tenants and field hands. Employers in Mexicali had a captive labor force since Chinese could not legally enter the United States. Anti-Chinese riots in other states compelled the Chinese to remain in the Mexicali Valley. Chinese recognized their importance to the profitability of the cotton industry, and they frequently went on strike to increase wages.[7]

As labor-intensive vegetables and cotton increased in the Imperial Valley, the dilemma facing farmers involved a cyclical labor market, the desire to keep wages low, and an earnest intention to prevent strikes. Since the labor supply was tight and urgent during harvests, migrants had strong bargaining power in threatening to walk off the job unless farmers raised wages. African Americans, for instance, took advantage of this situation. One farmer complained when eighteen "colored" men deserted him after another farmer offered "inducements." He lamented, "I have always paid my pickers 2 cents per pound and I suppose the other fellow came along and offered more money. I don't think this is a fair deal, as we farmers always upheld this figure and somebody is doing the dirty work."[8] Facing a dearth of workers, employers recruited African Americans from the South. An Imperial County newspaper editor boasted about the county's growing cotton industry in *Sunset Magazine*. The "care of cotton does not require skilled labor, and as to the harvesting it is often true that a small child can pick more of the white crowns in a day than a grown man." The author warned against hiring Japanese because they demanded high wages and went on strike. Instead, he argued, whites should recreate the racialized world of the plantation South: "The picturesque Negro, with his music and drollery was missed by the old-time cotton growers, but it is thought that they may be brought into the valley in the future."[9]

Despite the wish of some southerners to recreate the South in the California desert, other settlers vigorously protested the importation of blacks, and racism was rampant. Newspapers referred to them as "Sambos," "darkies," "coons," "black monsters," and the occasional "nigger." All racial groups loathed blacks, and some employers stopped recruiting them. When farmers advertised for Texas cotton pickers, for instance, they expressly mentioned hiring only Mexicans. Blacks began moving to towns where they worked as chauffeurs, realtors, cooks, truckers, carpenters, domestic workers, and manual laborers. For a short period, the sheriff's office employed a black officer.[10]

The shift to labor-intensive crops gave Japanese growers and shippers the opportunity to hire their countrymen. An early local history mentioned Japanese packed 1,000 crates of cantaloupes a day and a newspaper stated: "Large numbers of Japanese laborers are arriving daily to assist their countrymen in the cantaloupe business." Locals estimated "400 Japs will be brought to the valley to harvest the melon crop."[11] A few years later, "One hundred and eighteen pieces of baggage—the round sailor bag style—were put off at Brawley this morning from the Los Angeles train and almost that many Japanese alighted from the coaches, though there was a good sprinkling of [white] Americans." As expert melon pickers, the men came "ready to attack the fields." Hiring different racial groups caused tensions. In 1915, Japanese workers in a field who threw bottles at a white man were charged with assault with a deadly weapon.[12]

Filipinos were another important migratory labor force. After the government banned Japanese laborers, employers turned to Filipinos for field hands. When Congress halted South Asian immigration in 1917, farmers recruited more Filipinos. In 1920, the census counted 5,603 on the mainland, and ten years later there were 30,470 mostly young, single, uneducated male Filipinos in California. Filipinos were racially classified as "Malay," and they experienced withering racial discrimination and social ostracism. Crews of Filipinos first arrived in the Imperial Valley in 1916 to pick cotton.[13] Even though many employers refused to hire them because of a reputation for striking and fighting, labor contractors brought Filipinos to the valley to weed and to harvest vegetables. Mexicans, however, objected to Filipinos, accusing them of underbidding for jobs and aggressively pursuing Mexican women.[14]

During the Mexican Revolution (1910–20) over one million Mexicans crossed and recrossed the border, fleeing bloodshed, starvation, and political turmoil.[15] Braceros made significantly higher wages in the United States. Laborers in Mexico earned 15 cents a day and miners took home 50 cents a day, while in the United States, they made $1 to $3 a day and miners earned $4.50 to $5.50 a day.[16] The Imperial-Mexicali Valley's isolation made it difficult for Mexicans to travel to the region. Few Mexicans took the railroad from Texas to California, and braceros in the mining districts of Baja California had to take steamships to Tijuana, cross the border into San Diego, and then travel to the valley. Esteban Cantú, the governor of the northern district of Baja California, eased the arduous journey when he built a dirt road from Tijuana. Once in the Mexicali Valley, migrants easily crossed the invisible border into America since the Bureau of Immigration only had a few inspectors along the US-Mexico borderland before 1924.[17] Labor contractors took braceros to the desert during picking season. During labor shortages the Southern Pacific Railroad sold tickets from Los Angeles to the valley at a discounted fare.[18] Employers desired Mexicans because they met all the criteria for what they believed were a controllable workforce: they were easily recruited and discharged after the harvest, they worked for lower wages than other groups, and few desired to become citizens. To employers, Mexicans were inexpensive, mobile, unorganized, uneducated, impoverished agriculturists who "naturally withstood the heat." Mexicans were also desirable because they immigrated with families who worked in the fields.[19]

World War I Labor Shortage

The wartime demand for cotton in the Imperial-Mexicali Valley coincided with newly passed stringent immigration legislation. Employers responded to the paucity of labor by urging Congress to relax immigration restrictions for Mexicans. Agriculturalists used racial arguments to support unfettered Mexican immigration. In congressional hearings, farm leaders claimed whites were incapable of working in the boiling sun and only Mexican "peons" could survive in the desert. In May 1917, the secretary of labor suspended immigration restrictions for Mexicans for the duration of the war, but only permitted entry for six months.[20] Despite the removal of border crossing restrictions, few braceros showed up during the war.

Imperial Valley farmers had the opportunity to use American-born white workers but declined. In August 1918, the Half Century Association (HCA) of Los Angeles suggested supplying up to 2,500 white male workers for the upcoming harvests. Two sticking points arose during negotiations. First, workers demanded to be paid by the hour, but farmers insisted on a daily wage or piece rate (pay according to the work done). Second, HCA required employers to provide camp housing and transportation to the field. Landowners aiming to maximize profits refused to build decent housing. HCA's president made his case by appealing to racial solidarity: "This is a white man's country, why not keep it so?" He argued "aliens" should leave the state to open up jobs for whites. Predictably, growers declined the HCA offer.[21] Instead of embracing white workers, growers lobbied the government to create a surplus labor force. In September 1918, the Farmers' Labor Employment Association and the Employment Service paid Texans to travel to the Imperial Valley. Five months later federal officials complained about labor saturation, which resulted in workers making a miserable $50–$65 a month for a nine-hour day. When workers avoided the desert the following year, Japanese farmers banded together and placed labor advertisements in California newspapers. Japanese preferred to hire their countrymen, but noted they would hire Mexicans. Desperate for field hands, they promised to pay the astronomical wage of $300 a month.[22]

After the war, controversy in Mexico over the importation of Asians to the Mexicali Valley presented a potential solution to the Imperial Valley's perennial labor shortage problem. In 1919, unemployed workers complained about Governor Esteban Cantú's support for Asian immigration. The governor moved to contain the unrest by banning Japanese from entering the district, a decree that violated Mexico's treaty with Japan. Cantú's efforts failed to placate the masses. As dissatisfaction spread, in September troops in Algodones mutinied and killed four officers. Soldiers hunted down the deserters, and after a trial soldiers executed three ringleaders. A few days later, Cantú informed the secretary of state of his decision to suspend Asian immigration to the northern district until new federal immigration legislation was in place. Then Cantú released a letter to the public on September 16 of his decision and warned that a xenophobic rebellion would offer the United States a pretext to annex the peninsula.[23] The governor's ban stranded 320 Chinese in San Francisco. The Colorado River Land Company informed the federal government of the situation, and Cantú finally relented. Cantú's handling of the crisis diverted blame for his pro-Chinese migration policy onto the federal government, while the governor continued to grant special requests.[24]

Months before the mutiny at Algodones, Governor Cantú and cotton industrialists responded to the growing resentment against Chinese by forming the National Chamber of Agriculture of the district to recruit Mexican workers. The Agricultural Chamber charged $1 per cotton bale to pay for transportation expenses. Arnulfo Liera, a local merchant who became the president of the Chamber of Agriculture, had launched a company two years earlier with three steamships to bring braceros from Sonora and Sinaloa to Baja California and then truck them to Mexicali.[25] The Colorado River Land Company acted to minimize Mexican settlement by evicting migrants after the picking season and burning their brush shelters. Braceros either returned home at their own expense or crossed into the United States. The company's goal to banish unemployed cotton pickers coincided with the Imperial Valley farmers' desire to hire them, benefiting employers on both sides of the Imperial-Mexicali Valley. Viewed from this vantage point, Harry Chandler's vigorous effort to unite Arizona and Imperial Valley into a regional cotton exchange (discussed in chapter 3) was a strategy to create a superstructure of agro-industrial control. Thus Mexicali employers' recruitment of braceros to replace the Chinese set in motion Mexican circular transnational migration into California and back to Baja California. Over the next decade, braceros developed a labor network, starting in Mexicali and then moving up the spine of California and back in an oblong loop.[26]

Mexican Migrants

The Colorado River Land Company's importation of Mexicans set in motion the Imperial Valley's third labor market. In April 1920, Imperial Valley farm and business leaders met to eliminate the labor shortage problem once and for all by forming the Cotton Growers' Association to harvest the county's gigantic 122,000 acre cotton crop. At the first meeting, participants expressed interest in recruiting labor from Texas or Mexico. Farmers pledged to pay for the association's overhead and transportation costs. The association set wages, mediated pay disputes, and forbade bidding for workers.[27] Arizona cotton brokers and buyers, who financed many Imperial Valley farmers, encouraged them to start an organization modeled after Arizona cotton associations. An Arizona cotton representative suggested cooperation between Arizona and the Imperial Valley "to obtain labor for all these valleys, thus regulating the expense." Two hundred growers signed up when they learned Arizona pickers earned half the wages paid in Imperial County, potentially saving $3 million to $4 million.[28]

The cotton association advertised its intentions and lobbied the US government for an exemption to the ban on immigrant contract labor. "The workers are imported under contract, and if they refuse or fail to work on the ranches where they are placed, they are turned over to the United States officials for deportation. If they leave, their pictures are in the hands of officers, and they are sought and deported if captured. If they prove unruly or strike, they are deported." H. H. Clark, the general manager of the Colorado River Land Company, helped organize the Imperial Valley cotton association. He modeled it after the Mexicali cotton picker recruitment agency created after the government banned Chinese transmigration the previous year. Representatives from the Cotton Growers' Association called a meeting with Arizona agriculturalists, a US immigration official, and the Mexicali labor recruiter agency. Growers asked the Bureau of Immigration for an exception to the contract labor law. Agriculturalists also petitioned Governor Esteban Cantú to lobby the Mexican government for cooperation.[29] After negotiations with Mexico and the US State Department, the cotton association leaders required farmers to pay a $20 bond for each picker they needed. Employers would deduct a small portion from a worker's pay each week to repay the bond. When farmers objected to the fee, cotton ginners agreed to charge $2 per cotton bale to cover the association's operating expenses as well as "the construction of concentration camps, the sending of men into Mexico to get the laborers, the employment of men to keep track of them after they are here so that they will be certainly returned" after the harvest. Immediately, employers cut wages 25 percent. By late September, 896 workers came from Mexico, but seventy-seven were classified as "undesirables" (most likely they refused to accept the working and living conditions set by farmers) and deported.[30]

As cotton prices plummeted in a glutted market, the cotton association imported more pickers to drive wages down further. By mid-October, 1,400 pickers had been brought in and more were on the way. Halfway through the harvest, prices bottomed out, demoralizing growers, yet picking continued only because of low wages. Soon, the Cotton Growers' Association went bankrupt when farmers were unable or unwilling to pay the bale tax. With the cotton association defunct, laborers were left stranded. Some employers found the silver lining in the situation. J. E. O'Neil, a leading grower, commented, "You can take it from me that the growers this year have saved $15 a bale on cotton picking" using the cotton association. "I do not think it is so much the number of Mexicans brought in as it was the moral effect their bringing had upon the pickers already here," who were demanding the old wage scale. After the "imported help [arrived], these home pickers clammored [*sic*] for the work at any price." O'Neil expressed the farmers' perspective: "All of them are making good money at the present figures."[31] Although the cotton association collapsed in the postwar recession, its achievement importing large numbers of Mexicans followed two decades of labor shortages and the experimentation with a half-dozen ethnic and racial groups. For farmers, transnational braceros solved the valley's harvest labor shortage problem and lowered wages. In sum, the cotton association instituted Mexican circular migration on the California–Baja California borderlands, and, for the first time, gave desert farmers the upper hand over their workers. The campaign to import braceros transformed the field workforce and solidified a racially hierarchical agricultural workforce.

In 1921, the Mexican government renegotiated its treaty with China and ended Chinese transnational labor migration. As a result, Mexicali Valley farmers stepped up the importation of braceros. In 1923, for example, growers in the Imperial-Mexicali Valley advertised for workers from Sonora and Sinaloa to be transported by steamship. Between 1921 and 1924, 13,000 Mexicans arrived in Mexicali. Farmers in Imperial County hired two-thirds of them. Within five years, 24,000 Mexicans arrived in Mexicali. The Colorado River Land Company, the region's primary employer, spent $400,000 from 1923 to 1925 to transport the braceros. Although Mexicali landowners hired braceros, Chinese, Japanese and South Asian tenants lived year-round on the ranches.[32]

Segmented Labor in the 1920s

The scope and complexity of the perishable fruit and vegetable industry made recruiting and retaining labor difficult. Employers claimed labor shortages to convince the US and Mexican governments to grant permission to import thousands of Mexicans. By 1920, they formed the majority of agricultural workers in California and a third of Imperial County's population. With a plentiful supply of braceros, employers created a dual labor market by segmenting workers in shed and field jobs according to race and gender. Japanese and white men held most shed jobs (except for female tomato packers), while Mexican and Filipino men worked in the fields. After the war, farmers were in maximum-production mode. It took an enormous amount of labor to level, plant, weed, cultivate, pick, haul, pack, load, and ice 9,000 railroad cars of produce in three months. Resident workers planted and irrigated dozens of crops, but farmers needed 10,000 harvesters for the fall and winter harvest. In addition to migrants, employers recruited school children, town folk, drifters, and coastal residents.[33]

A seasonal description of harvesting labor-intensive crops illustrates how farmers coordinated workers during the picking season and how racial stereotypes influenced hiring decisions. Lettuce planting occurred in mid-September and lasted for seven weeks. Cultivation and thinning took place from early November to Christmas. Employers hired 3,000 men just to pull weeds in the lettuce fields. This "stoop" labor required nimbleness and quickness, and

Multiethnic laborers harvesting carrots in open-air factory. Courtesy 3A18-24, Imperial County Historical Society, Pioneers Museum, Imperial, California.

employers attested that Filipinos performed this task the best. It took about 7,500 harvesters to bring in the lettuce crop, which began the first week of December, peaked in late January, and ended at the end of April. Lettuce was extremely delicate, and if the weather and prices cooperated, it brought in a fortune. Peas, grown in the north end of the valley, were the second most important fall and winter crop. The fall pea harvest lasted until late December. Spring peas were planted once the frost subsided. Picking started in February and ended in early April. On the other hand, melons (cantaloupes, watermelons, honeydews, and honeyballs) were "heavy" produce, which farmers grew at different times throughout the year to avoid saturating the market. Some growers planted early melons, banking on a premium price, while praying frost, bugs, or fungus did not ruin the seedlings. In May and June, when the temperature hovered over 100 degrees, about 7,000 workers harvested the valley's largest melon crop.[34]

Labor and Racial Attitudes in the 1920s

Imperial Valley's racial and class structures testified to dominant racial ideologies. A report released after the war of child cotton pickers obliterated simplistic notions of race and class ordering. Mexicans represented a majority of cotton pickers, but black and South Asian cotton farmers also employed black harvesters as well as Mexican and white children. Comments by agriculturalists shed light on the economics of large enterprises and how racial stereotypes shaped labor hiring decisions in favor of Mexicans. In the past, W. S. Fawcett, a large shipper, had experimented with whites, Filipinos, and Japanese, yet he found them all lacking. Japanese picked green melons, whites hated to hoe and could not take the heat, and Filipinos were "not as reliable"—code for a willingness to walk off the job when insulted or shorted on payday. Another farm leader remarked, "Mexicans have to be supervised and watched," but they "are better than Japs and more trustworthy than Filipinos."[35]

Racial and cultural attitudes colored how whites viewed Mexicans. According to historian Gilbert González, white Americans classified Mexicans as "peons," a racial-cultural term packed with a host of meanings. The unflattering term projected an image of docile, meek, stupid, backward, lazy, indolent, and shrouded in a "mañana" culture, or unable to plan for or work toward the future. One agribusiness leader typified this mentality when he lauded Japanese and disparaged Mexicans: "It costs more to produce stuff when you use Mexican labor because they are inefficient labor and then they just can't produce as much as the Japanese." He believed the Japanese were "especially adapted to do that tedious sort of work." In contrast, "Mexicans are poor laborers and they will only work until they get a little money. When a Mexican gets his pay on Saturday night, he won't work any more until he has spent it all and is broke."[36] Congressional testimony and agricultural publications attested to the docility and tractability of Mexicans.[37] A "leading" archaeologist of Mexico offered a racial thesis for the inefficiency of Mexican labor, claiming Indian genetic ancestry explained their laziness, fear of machinery, addiction to alcohol and gambling, and weakness due to widespread venereal disease.[38]

At the end of the day, racial stereotypes were a smoke screen to justify paying Mexicans paltry wages. The Imperial County district attorney stated it plainly: "I know it isn't easy to secure white labor, but these men don't want to pay white men's wages. They get a lot [of workers] in Mexico and after they have been here a while they get an American education" and demand higher wages, so "the growers get some new Mexicans who are satisfied to work" for lower pay.[39] One grower remarked, "Mexican labor is good labor, and we couldn't get any other class of labor for anything like the same money." A company executive commented: "Large-scale production would be impossible without" Mexicans for "costs would be increased 50 per cent."[40] Willingness to live an itinerant lifestyle added another bonus. Workers who left after the harvest relieved the community of providing public assistance for the unemployed. The secretary of the El Centro Chamber of Commerce emphasized this point: "We need the labor of the Mexicans. They are not like the Japs and Hindus. They don't come to stay. They are satisfied to labor."[41]

Employers exploited racial discord whenever possible by pitting groups against each other to inflame racial antagonism, discourage labor unity, and encourage ethnic and racial competition for jobs. Juan Estrada, a labor leader in the valley, disparaged Filipinos for lowering wages. Braceros had families to support and some had home mortgages. Young, single Filipino men didn't have these expenses and they could afford to underbid Mexicans.[42] At best, racialized groups rarely interacted. They battled each other for jobs and jousted verbally and physically in the fields and in towns. Many Mexicans despised Japanese farmers, claiming that they were disrespectful and shorted them out of pay. Mexicans considered South Asians and Filipinos sexual competitors, since both groups pursued Mexican women for wives. Furthermore, South Asian and Japanese farmers had a different class perspective from their Mexican workers.[43]

Education and Field Labor

State-mandated education for children profoundly impacted the workforce. Recruiting Mexicans ushered in a new era in Imperial County because, other than whites, they were the only large racialized group with families. State officials looked the other way when adult workers lived in cars, tents, and shacks, but authorities zeroed in on housing and schooling for children. A 1905 California law required children between six and sixteen to attend school full time. During the first decades of colonization, farmers and parents frequently ignored the school attendance law. In 1918, for example, the *Brawley News* reminded parents of compulsory school attendance. Officials estimated five hundred children of school age were working in the fields and absent from school. The truant official sighed, "When the employers are warned concerning children in the fields, the cotton pickers simply move into another district" to evade truant officers.[44]

After the war, school attendance officials estimated between 2,500 and 3,000 white, Mexican, and black children picked cotton. When the 1920 cotton exchange went bankrupt, it caused widespread truancy and starvation among the bonded Mexicans whom it left stranded. Charities helped with the direst cases while migrants scrapped by for months to save money to return home since they could not legally stay in the United States. The cotton debacle turned many settlers against future proposals to import temporary Mexican migrant families.[45] In the early 1920s farmers lowered wages paid to workers by hiring children. The school truant officers tried their best to punish farmers for letting children work during school hours. In one case, Kakoo Singh and Atma Singh were arrested for employing children in the cotton fields. A disgusted school principal told an investigator, "These employers are like children, just boys grown up, and will keep on breaking the law until somebody catches them. They say it will ruin their business if children do not work. I suppose they mean by that they will have to raise the wages of adults." A trustee zeroed in on why Mexican parents put children to work in the fields: "The Mexicans are the last people on earth to ask for charity—they will starve before they will ask for food—but they work their children."[46] As the state's regulatory apparatus grew in the 1920s, state and local officials pushed for enforcement of school attendance laws. Newspapers printed the offenses and names of farmers, foremen, and parents for abetting truancy. Fines and public humiliation contributed to the crackdown on child labor. By 1927, the supervisor of county schools reported absolute observance of the mandatory school attendance law. School attendance records confirmed the shift from multiracial family labor to single adult male transnational migrants of color.[47]

Passbook Plan 1926

A wave of racism, xenophobia, and eugenic thought swept the nation during the 1920s. Nativists railed against undesirable immigrants and racial mixing. In 1924, Congress lowered immigration quotas, raised entry costs to $10 for a visa plus $8 for a head tax, ended Asian immigration, and created the Border Patrol. Employers, however, persuaded Congress to exempt the western hemisphere. Prior to the Border Patrol, there were very few immigration agents patrolling the US-Mexico border. The Immigration Service and undocumented Mexicans did not take illegal entry seriously because agents operated under orders to facilitate border crossing, and illegal entry was a misdemeanor until 1929.[48] Cotton farmers grew alarmed as more agents appeared along the desert sector detaining unsanctioned crossers, and they organized a cotton exchange to import Mexican pickers. Other ranchers and labor contractors blocked the effort, pointing out the previous year the labor surplus left many families penniless at the end of the season and dependent on public assistance, costing taxpayers thousands of dollars.[49]

Despite the setback, cotton farmers were satisfied with the glutted labor market, for an estimated 70–75 percent of the Mexicans in the valley had entered the United States illegally. The Border Patrol hired more agents in 1925 and began capturing undocumented border crossers. A farm spokesperson testified in a congressional hearing that "the Mexicans were like a lot of jack rabbits that were being hunted." Border Patrol agents released Mexicans into Mexicali and the next day they clandestinely entered the United States. Concerned farmers petitioned Congress and the Border Patrol to end the harassment.[50] The following year, employers set up a countywide labor exchange under farmer control to end deportations and eliminate competition for harvesters, which allowed workers to demand high wages. Building upon the experience observed from the Mexicali Chamber of Commerce labor bureau, in March 1926, the Imperial Valley Chamber of Commerce formed the Associated Labor Bureau. After extensive discussions, the local Border Patrol chief agreed to a program called the "passbook plan" to let undocumented Mexicans make installment payments on the $18 immigration fees at local banks. Once paid, the individual took the literacy test and obtained consular paperwork and photos. It was a local arrangement with no validity elsewhere. In this way, the Imperial Valley established its own immigration policy with the government's unofficial blessing. Farmers now controlled thousands of braceros who were subject to deportation if they left the county.[51]

Agricultural leaders tapped Chester Moore, a Spanish-speaking long-time Calexico banker, to manage the Labor Bureau. Moore registered over three thousand persons in two months. Immigration officials also waived the immigration fee for seven hundred Mexicans for prior payment or for living in the United States for the previous five years. The bureau registered 6,217 undocumented Mexicans from April 1926 to June 1927. After the Mexican consul reported abuses, the Mexican government filed a complaint, leading to an investigation by state labor statistician Dr. Louis Bloch. He discovered no labor shortage existed at that time or the year before. Bloch concluded the labor shortage was a ploy to lower wages. In his report, Bloch speculated the bureau's deal with the Border Patrol would allow banks and employers to defraud braceros by illegally collecting immigration fees.[52]

Shed Labor

Agribusiness had a firm hand on the field labor situation; what gave employers fits were white packing shed workers. They were the barons of labor in the desert, since packing jobs were respectful and well paying. White men packed "heavy" produce like cantaloupes, and white women sorted and packed tomatoes and other lightweight produce. In July 1920, for instance, a newspaper column noted: "The grape girls will be with us once again" to pick and pack near Imperial and El Centro. This particular group of "girls" numbered seventeen and resided near Fresno, California, and local white women planned to join them.[53] Before the war, the Japanese dominated produce packing, but during World War I, shippers built packing sheds and hired white packers, known as fruit tramps. Concurrently, community members pressured Japanese packinghouses to hire white shed laborers. Once white shed workers made inroads into the packing sheds they formed trade unions and vigilantly protected their jobs against Japanese and Mexican rivals. During the 1920 Alien Land Law campaign, a shed union passed two resolutions with virtually identical language. One supported the Alien Land Law and the other opposed the Cotton Growers' Association's Mexican worker importation program.[54]

The diversity of shed labor is seen in the 1922 cantaloupe harvest. Employers recruited five hundred Japanese packers, as well as hundreds of white "fruit tramp" packers and box makers. Shed nobility fancied itself as racially and economically superior to field laborers, with farm leaders encouraging and exploiting white racial chauvinism. "White packers are used because a high degree of intelligence is required for this work," remarked a vegetable manager. "Also race prejudice is a factor. Introduction of Mexicans in work commonly performed by whites would lead to serious disorders between the two groups."[55] A fieldworker remarked, "Employers try to keep workers from [organizing] by keeping them fighting each other." Packers looked down on Mexicans, and Mexicans were resentful because whites failed to support them during a strike.[56] Expressing the valley's racial rules, a shipper revealed employers hired whites as a matter of "civic responsibility."[57]

Labor Contractors, Wages, and Housing

During the 1920s, agriculturalists pursued numerous strategies to maximize profits. Until 1920, wages fluctuated according to economic conditions, labor supply, and the crop, but overall they remained relatively high due to the level of difficulty, harsh weather, and labor scarcity.[58] Labor scarcity dissipated with the importation of braceros by Mexicali Valley landowners. In the 1920s, California paid the highest wages in the nation for agricultural workers, but it was one of the most expensive states to live in. Gas, vehicle maintenance, rent in the primary home, and unemployed downtime drained meager resources. Laborers lived constantly on the edge of devastation due to sickness, injury, a broken vehicle, or a wage cut. Few workers could afford homes, medical care, or even clothes and shoes for their children. A Mexican field hand said with members working together the family made "pretty good" wages. But after a harvest, gas and food quickly drained savings while waiting for jobs. In anguish, he said it was unbearable to hear his hungry children crying for food.[59] Integrated agribusiness was in an echo chamber, speaking and hearing the same message from elites such as R. L. Adams, the director of the Giannini Foundation of Agricultural Economics at the University of California, Berkeley, who published a popular farm management textbook filled with negative stereotypes of fieldworkers and advised farmers to keep wages low and to dispense with camp housing.[60] Occasionally, employers received more than they bargained for when they demanded state assistance. In 1926, for instance, the state opened an employment office after farmers demanded assistance to put men to work. The labor commissioner found jobs for the unemployed and also prevented labor abuses. In November of that year, a labor commissioner filed 193 labor claims. In response to a spike in labor abuses, in 1928 the legislature passed a law requiring employers to pay workers at least twice a month. If room and board was provided, workers could be paid once a month. Imperial County labor claims plummeted from $5,000 to $300 in one year.[61]

Migrant workers had very little control over their wages and working conditions, and the system was rife with the opportunity to exploit, which, when exploitation occurred, plunged workers and their families into an economic abyss. Unpaid wages were workers' most common complaint. Growers often signed a contract with a labor contractor to avoid the arduous task of recruiting, paying, feeding, transporting, overseeing, and paying the medical bills and workman's compensation of laborers. Since state law did not regulate contract laborers, it was common for farmers and labor contractors to either deny pay or to short workers' wages. When farmers or labor contractors absconded or refused to pay, workers could lodge complaints with the labor commissioner.[62] A standard labor contract stipulated a set amount for every crate or sack harvested, and the farmer paid the contractor weekly. Farmers frequently received loans from shippers or landowners, and the loans involved contracts that set field labor wages. Contracts frequently determined workers should earn $3 a day, but farmers and labor contractors often skimmed money from workers' pay. Contractors hired workers for $2.50 a day and pocketed (or shared with the tenant or farm manager) the rest. Contractors frequently charged exorbitant fees for transportation, room, and board. They also underpaid workers and kept the difference.[63]

A major point of contention was the "bonus" or "holdback" clause of a labor contract. A provision allowed the grower to withhold 25 percent of the wages as a retainer to make sure the crew or individual finished the harvest. If a grower lost money in the harvest he could not pay the last round of wages, which included the bonus. The state could not squeeze unpaid wages out of a broke tenant, and a laborer could not sue on the basis of the crop under the mechanic's lien law. Even more flagrant, dastardly contractors sometimes absconded with the last week's paychecks, including bonuses. Many complaints were lodged against packing sheds that paid corrupt labor contractors. Sheds refused responsibility and workers had little recourse to obtain their pay.[64] After a strike in the desert, state statistician Dr. Louis Bloch investigated the labor contractor system, noting it "saves the grower the trouble of hiring each worker and of keeping a record of his account." Yet the contract labor system bred opportunities for abuse. Bloch concluded, "These defalcations are not infrequent and the Mexican laborers in the Imperial Valley have suffered considerably" because of fraud.[65]

Unpaid salaries, low wages, and unemployment between jobs resulted in poverty. Employers added to dire living conditions when they ignored the law requiring sanitary housing for fieldworkers. Many migratory workers could not find rentals in town, and they were forced to lash together wooden and cardboard shacks without electricity, accompanied by fly-infested outhouses and water hauled from community pumps or irrigation ditches.[66] An outside observer commented on Brawley's eastside, "homes are hovels; few have lawns; there are few or no sanitary conveniences; and few homes have running water."[67] Housing for whites on Brawley's west side had electricity, running water, plumbing, and graveled or paved streets. To justify the disparity, employers claimed Mexicans enjoyed substandard living conditions. A leading agriculturalist, for instance, dismissed Mexicans as filthy animals who did not appreciate modern housing.[68]

Boosters and conscientious residents fought an uphill battle to drag the county into the twentieth century. One official complained: "The housing people would like to do something about our towns but they can't and they won't get much support from our judges" and other county officials, who were growers or absentee landowners. He stated farmers "may get fines and penalties but they will be small and cheaper than houses."[69] Urban residents took preliminary steps to improve living conditions. People of color petitioned city officials to get regular trash pickup, provide utilities, and to beautify neighborhoods. Fearing epidemics, county medical personnel tried to improve personal hygiene and residential sanitation by quarantining cases of tuberculosis and smallpox.[70] Deplorable housing led "to friction with state labor, sanitary and housing authorities, and presented other objectionable features," wrote an agribusiness organ. Ultimately though, county officials and employers refused to build adequate housing, resulting in illness and deaths due to exposure and unsanitary conditions.[71] Housing in the countryside was more deplorable than in town. Absentee landlords refused to build homes for tenants or workers. Tenants transported shacks to new fields every three years. Small farmers built ramshackle homes and employers ignored the Interstate Commerce law mandating housing for out-of-state workers. Farmworkers pitched tents or slept under trees and in automobiles. Some employers let workers build huts made of discarded wood, cardboard, tin siding, and brush. A state investigator reported: "Health and decency under such living conditions are inconceivable."[72] State officials exhorted recalcitrant employers to provide sanitary housing and had to endure decades of pathetic excuses. In a typical understatement, an official attributed the lack of workers "to the fact that labor was not treated in a manner that invited permanency."[73]

In 1926, Fred Rugg, a state migrant camp inspector, visited 150 labor camps and found none in compliance with sanitary regulations. Rugg prepared criminal charges against camp owners. Predictably, local judges defended the growers. Five years later, the director of the Department of Industrial Relations cited Imperial County camps as among the worst in the state. When labor commissioners visited three hundred field camps during one winter, they wrote: "we found filth, squalor, and entire absence of sanitation, and a crowding of human beings into totally inadequate tents or crude structures built of boards, weeds, and anything that was found at hand to give a pitiful semblance of a home at its worst. Words cannot describe some of the conditions we saw." Investigators could only imagine summertime heat, with flies and families roasting in tents and shacks.[74] After inspecting work camps in the valley in 1934, a traumatized federal labor investigator wrote to his superior: "If you could see the conditions there you wouldn't sleep for nights." In another heart-wrenching case, a social worker stumbled on an abandoned schoolhouse where fifty Filipinos had camped over the winter, collectively earning less than $50. Ineligible for public assistance because they were migrants, and frequently assaulted by whites, the gaunt men subsided on handfuls of rice and bread.[75] During the Great Depression, when federal officials threatened to enforce housing and sanitation laws, employers rushed to make cosmetic improvements. California officials strenuously opposed federal involvement, and compliant state officials who inspected the desert's labor camps reported splendid progress. Employers abandoned the worst sites and replaced about "200 bush hovels" that sheltered over six hundred people with tents, 125 new toilets, kitchen houses, and garbage pits.[76]

Industrial farmers strove to maintain power and control over workers by keeping them in a subservient status. In addition to recruiting a multiracial labor force to keep workers divided, agribusiness resisted every effort to pay fieldworkers a living wage and to provide sanitary housing as required by law. The establishment of the labor contract system in the 1920s allowed farmers to avoid responsibility for payment of wages, workman's compensation, and housing. By the end of the decade, patterns in labor relations showed how farm leaders subjugated workers in the same way they "collared" the Colorado River, shaped the landscape, and eradicated wildlife.

The Great Depression

The Depression ushered in the fourth period in the evolution of the valley's agricultural labor structure and worsened the working and living conditions of farmworkers. Between 1929 and 1935, some 300,000 train-hopping "hoboes" and caravans of displaced unemployed whites and smaller numbers of blacks barged into California's migratory labor circuit. The peak of American migration west during the 1930s occurred between 1935 and 1937. California whites racialized the destitute white migrants by colloquially calling them "Okies," "Arkies," "Texicans," and "white trash." Most were former tenants who had gone bankrupt, while others sought to escape the ravages of the Dust Bowl. These migrants came primarily from New Mexico, Oklahoma, Texas, Missouri, and Arkansas. In the Imperial Valley, they were cast down the racial ladder to the status of "colored" and were forced to live in the "Mexican colony" with their children attending school with children of color.[77] "Okies" stood the state's and the Imperial Valley's field labor pool on its head. Statewide, unemployed whites drove up the labor supply to a high of 185 harvesters for every 100 jobs in 1933. In cotton districts, wages of 50 cents a day were common.

It took $780 a year for a family to survive, yet the annual income for migrants averaged $350 to $400.[78] In Imperial County, wages averaged 30 cents an hour in 1920 and plummeted 12.5 cents in 1933. One Mexican worker said pooling income from family members was critical for survival: "Boys and girls can help pack grapes from about eight years of age on. They are an expense; after that they can help you." José Pérez exemplified the hardship. In 1939, he made 85 cents a day to support a wife and five children.[79]

The stampede of American migrants into California corresponded with the forced repatriation of over half a million unemployed Mexicans during the decade, and congressional hearings to put Mexico on the immigrant quota list. While Mexicans were pressured to repatriate, hundreds of thousands of braceros voluntarily returned home. Mexican leaders complained of the harsh treatment of expatriates in the United States, and they suggested government assistance to resettle them, including giving them farmland. For instance, in 1930 a leading Mexican functionary criticized foreign ownership of the Mexicali Valley and called for the deportation of Asians. Voluntary repatriation stepped up exponentially in response to the pronouncements by Mexican president Lázaro Cárdenas (1934–40) to return home and build up the nation. Cárdenas expropriated millions of acres from foreign landowners and resettled repatriates in massive irrigated farm projects. Across the continent in Washington, DC, in 1930, industrial spokesmen, including Chester Moore of the Western Growers, and Harry Chandler of the Mexicali Valley's Colorado River Land Company, urged a congressional committee not to put immigration quotas on Mexico. Farm leaders claimed the law would "bankrupt" growers in the Southwest and open the door to "Porto Ricans," Filipinos, and other "undesirable" races who posed a "social menace." Once again, testimony from irrigated desert agribusiness succeeded in maintaining a migration corridor to saturate the labor market and to disempower workers.[80]

The potent combination of American white and black migrants, starvation, and low crop prices muddled Imperial Valley's racial hierarchy and forced agribusiness to reexamine the race-class labor structure. For three decades, employers had treated white agricultural labor with respect. Working-class whites and employers shared a common interest in the economic vitality and stability of agribusiness and in maintaining white supremacy. White bosses and employees greeted each other on the streets, at ball games, in stores, and shared fellowship at fraternal clubs and the American Legion. For more than a decade, Mexicans had been the primary field labor force, and planters vigorously prohibited whites from toiling in menial field jobs in order to maintain racial and economic segregation and domination. Farmers recalculated the racial-class order as they faced foreclosure and heard desperate pleas by starving whites begging for jobs. Growers hired white tenants, pushing South Asians and Japanese tenants into migratory labor. Other employers hired whites and blacks to compete with Mexicans and Filipinos to exploit racial hostility, minimize labor unity, and drive down wages.[81]

In 1929, cotton growers in Yuma, Arizona, advertised for workers and promised high wages. Later, when white migrants drove to the Imperial Valley demanding jobs, growers grew indignant upon hearing about the high wages in Yuma and refused to hire them. Instead, farmers beseeched the government to allow five hundred Mexicans into the valley on contract and to waive the visa and head taxes, which the government declined. Some residents supported the Mexican importation request, while others became incensed when farmers would not hire destitute whites. When newspapers showed Imperial County was one of nine "white spots" of economic prosperity in the nation, impoverished whites streamed into the valley.[82] In February 1930, when a mass of railroad "hoboes" rushed into the Imperial Valley, sheriffs fired pistols to scare them off. As the economy worsened, American migrants camped along roads and under trees. Robberies, assaults, and murders increased dramatically. In previous years, police arrested or chased out undesirables, but the jails now overflowed with vagrants.[83] During the winter of 1933, a motorized army of 5,000 whites stormed into the desert. By the end of the decade, whites replaced Mexicans as the primary field hands in the Imperial Valley. Farmers hit the jackpot, announcing the need for abundant labor at a moment's notice and openly boasting "two or three men for each job was about right" during the picking season. Employers drew up a plan to hire men for a minimum of five hours a day for a minimum of $1.12 1/2. Mexicans and migrant whites lashed out at employers for offering "a starvation wage or less." When government officials confronted farm leaders about families camping alongside roads, employers replied the homeless enjoyed the desert's moderate winter climate.[84]

Dr. Paul Taylor, a labor economist at the University of California, Berkeley, and the foremost authority on farm labor in California, accepted a position with the state welfare agency to gather research on migratory workers. He crisscrossed Imperial County in 1935, jotting down a litany of complaints: horrible living conditions, low wages, and few good paying jobs. One white man commented, "The growers want the laborers to have to crawl on their bellies & come over to get work." A couple of fieldworkers expressed the vicissitudes of migratory life: "It[']s kind of scary—you work a while, & lay up a little, & then go broke, [and] then move" to the next harvest. Ruby Cassos recounted pulling carrots for two or three days and making 40 to 50 cents a day.[85] Taylor visited three hundred carrot "pullers" near Holtville. The motley crew consisted of blacks, Mexicans, Filipinos, and a man from India. One carrot picker explained the economics of agricultural labor. Pickers had to buy crates and ties. Crates cost $1.40 and carrot ties were 24 cents a pound. In four days a picker netted $4.95, but the meager wages provided dignity in an honest day's labor. "Even if we get only 50¢ we got to work," said one picker, because a strike would fail, since "we're all disorganized now." Resigned to the unbalanced power relationship, workers accepted the "starving wages."[86]

The proliferation of destitute whites overtaxed local charities, throwing the duty of social assistance to the state. Some white fieldworkers reflected ambivalence with state assistance. Free food was not only aid, it was also a handout, which violated the principals of what historian James Gregory called the "plain folk Americanism" of "Okie" culture. It celebrated individuality, self-sufficiency, patriotism, anticommunism, and a deep reservoir of racism. One Oklahoman enunciated the prevailing belief in individual responsibility. After picking peas for five days he made $5.66. He disagreed with the popularity of relief, believing welfare was bad for the work ethic. "This relief business is all fake anyway. When they get on it, they don['\]t want to work anymore. All I want is a chance to make an honest living," he said. By the late 1930s, farmers had become addicted to white fieldworkers.[87]

The New Deal

During the Progressive Era, California passed a host of social welfare and labor laws affecting agriculture, yet by the mid-1920s the commitment to social reform waned with a string of conservative Republican state governments. For instance, lawmakers pared back funding for the Immigration and Housing Commission, crippling its authority and its ability to investigate complaints of unpaid wages, substandard housing, and child labor. Furthermore, compliant judges refused to enforce laws benefiting farm laborers.[88] Residents on both sides of the border endured indescribable hardship. In California and in Baja California, local government did little to alleviate suffering and federal assistance was nonexistent.

Beginning in late 1930, secular and religious charities and the Red Cross in the Imperial Valley distributed food and clothing for the most heart-wrenching cases. County supervisors gave food and gasoline to destitute visitors on the condition they leave the region. Homeowners put unemployed job seekers to work cleaning homes and yards in return for food. To ensure residents took field jobs, police chiefs screened relief cases before dispensing charity. Town officials also put the unemployed to work on public projects in return for groceries. One program required food recipients to chop wood for a meal.[89] A year later, county officials contemplated a plan hatched by the governor for police to pick up "moochers" and force them to stay at the fairgrounds. From there they would be transported to a huge roadwork camp. Officials implemented the plan when large numbers of out-of-state migrants and roving bands of homeless men showed up. Between 1931 and 1933, the state set up 250 temporary labor camps, housing 15,000 men. In exchange for room and board the men worked on public projects. In Mexicali, after numerous people died of starvation in February 1931, women and children marched demanding food. As shopkeepers distributed food to prevent looting, three thousand braceros descended into town when farmers cut wages to $1.20 a day and laid off scores of pickers. Laborers threatened to loot the town if jobs did not materialize. Local authorities called for troops to guard grocery stores, while business owners dismissed American workers and replaced them with Mexicans.[90]

Immigrants were ineligible for public assistance by virtue of the "likely to be a public charge" stipulation of the immigration law. Once an immigrant asked for public assistance, or in some cases health care aid, they became eligible for deportation. To protect against this drastic measure, the Mexican consul in Calexico launched *mutualistas* (self-help associations) in the large Imperial Valley towns in 1925 to raise funds and assist needy expatriates. The next year, state labor statistician Louis Bloch told local farmers that contrary to popular opinion, "It is not his nature that makes the Mexican take charity, but the fact that he is underpaid. He has little share in the wealth[,] which he helps create. Considering what he produces, in spite of the fact that he may be the recipient of charity at times, he is a distinct asset, not a liability," for he produced far more than he took in charity. During the Great Depression, Mexican, South Asian, and Japanese immigrants cared for their own. The Mexican consul founded local chapters of la Cruz Azul Mexicana (the Mexican Blue Cross), a relief agency, to hold fund-raisers and dispense food and money to the unemployed. Similarly, Japanese hired unemployed members of their group and paid the medical bills of destitute compatriots.[91]

Examining charity in Brawley reveals the inner workings of private relief and the magnitude of the social and economic crisis. Relief organizations split the welfare load. Local charities helped families who lived in town for over ninety days. The Salvation Army assisted unmarried transients, and the Red Cross and American Legion aided veterans. Despite assistance, relief officials reported that in three months, seven people in Brawley died of malnutrition and related illnesses due to poverty. By the end of 1932, California was in economic straits. Within one year, 1.2 million of the state's almost six million residents accepted public assistance. In 1932, the amount of aid given to homeless and starving persons in Brawley was staggering. Two dozen charitable organizations assisted 2,000 families and fed male transients. An additional 515 male heads of households received public works jobs paid by charity. Relief volunteers provided medical assistance to 1,200 persons and gave out 3,300 articles of clothing. The Salvation Army carried most of the relief load, distributing 27,912 meals, 6,061 nights of lodging, and clothes and shoes to 1,007 single men. Additionally, it paid 1,469 men to work on city projects, and relief groups found 237 jobs for single men. Finally, 415 families received rent and grocery tickets.[92] As relief funding dried up in December 1932, Brawley's aid organizations merged to provide a uniform system of distribution. A month later, funding was on the verge of depletion. In an extraordinary acknowledgment of the crisis, in March 1933, the staunchly Republican *Brawley News* editorialized its support for "dictatorial" powers for newly elected Democratic president Franklin Roosevelt. By the summer, when self-help options collapsed, relief groups released a record of their activities. Brawley dispensed assistance to 54,000 people between November 1932 and May 1933; El Centro assisted 30,000; and Calipatria helped 4,000. Smaller towns assisted 6,000 people.[93]

Into the vortex of starvation and homelessness stepped President Franklin Roosevelt, who proposed an eclectic mix of economic and social government-directed reforms known as the New Deal. Congress established the Federal Emergency Relief Administration (FERA) to provide immediate direct relief to states on a matching basis. FERA gave $1 for every $3 the state spent on relief. The State Emergency Relief Administration (SERA) coordinated and dispensed aid, but FERA set policy. The Federal Transient Service (FTS) located jobs for the transient, interstate homeless. The FTS fed all unemployed transients, including strikers. The federal government also created the Work Progress Administration (WPA) and the Reconstruction Finance Corporation (RFC) to create make-work construction jobs. Unlike field labor, WPA jobs were "respectable," and unemployed blue-collar workers, farmers, merchants, and professionals applied for them. California set rules to benefit its residents over migrants and immigrants. Applicants for state relief and federal make-work jobs, for instance, had to have resided in the state for at least one year. RFC guidelines also excluded "aliens" who could not prove California residency for three years.[94] Despite the unequal relations of power in the Imperial Valley, fieldworkers rose up in a series of spectacular strikes. The efforts by workers to improve their pay and working conditions received national attention for police brutality and violations of civil liberties. Imperial County's brutal reputation gave it the ignoble distinction of being the only county in the state where the federal government controlled emergency relief. Agriculturalists launched a fierce offensive to regain control over relief decisions and to exclude strikers and American migrants from federal work programs.[95] In October 1933, after twenty-two years of lobbying, farmers in the desert hit pay dirt when the secretary of the interior allocated $6 million to begin digging the All-American Canal. Local officials locked out American migrants and "aliens" by restricting canal jobs to resident unemployed male citizens.[96]

That same month, an enormous cotton strike in the San Joaquin Valley ended in a partial victory for Mexican and white pickers and communist union organizers. The federal government played a role in the labor dispute by feeding strikers. The union's success shook California agribusiness to the core. The federal government directly challenged agribusiness's hold over workers when the Civil Works Administration (CWA), a federal construction program, began accepting applications from all aid recipients, including workers on strike. After a meeting with Imperial Valley farm leaders, the head of the local Relief Commission announced the denial of aid to workers who refused field jobs. The new policy had an immediate impact, claimed a large pea grower, who credited the Relief Commission's decision for increasing the number of workers by 35 percent. During the winter, the federal government poured more money into the county, and the Relief Commission designed projects specifically for residents and nonresidents to ensure the selection of locals for the best jobs.[97]

The year 1934 was a turning point in agribusiness-labor relations in the Imperial Valley. During a January pea strike, the Civil Works Administration announced job cutbacks. Authorities urged the CWA to reconsider its decision, as growers did not want to put local unemployed whites in the fields. Racial privileges combined with a spirit of entitlement pervaded the county's leaders. Agriculturalists wanted the government to assist unemployed white residents while simultaneously helping farmers break the strike. As police and vigilante farmers beat and arrested picketers, federal officials fed strikers. A newspaper editorial during the pea strike expressed the frustration of the farming community: "The irony of an economic system which permits people to go hungry while food is wasting in the fields is nowhere more strikingly illustrated than at Calipatria this week. The pea crop goes unharvested, while down the road a short distance federal agencies" fed strikers. Meanwhile, "the growers who are representative of the class which through taxation must furnish the money to feed the idle workers, are losing their" investment. "Figure it out if you can—it's too complicated for me."[98] In mid-February, in the midst of the pea strike, there were 1,892 relief cases. Mexicans, whites, and "colored" folk congregated on the streets of Calipatria waiting for groceries. As strikers received food, the Civil Works Administration employed 1,185 mostly white residents. Although farmers lost the argument to deny food to strikers, it shaped policies benefiting residents with jobs while giving unemployed migrants food between harvest jobs. Before the cantaloupe harvest in 1934, for example, the government fed 9,895 persons. Out of a total of 1,932 families receiving groceries, only 678 families were white. Most of the aid recipients were Mexican. The unemployment crisis for local whites subsided in October when work began on the All-American Canal. After a fierce debate, the government relented to agribusiness demands to put 2,000 resident farmers to work digging the canal with mule teams and fresno scrapers.[99]

The 1934 election presented Democrats with the opportunity to end four decades of Republican domination. Socialist Upton Sinclair changed his party affiliation to Democrat and became the party's nominee for governor. Republicans rallied around Governor Frank Merriam and smeared Sinclair as an enemy of capitalism and a supporter of socialist policies. The unprecedented assault by the state's newspapers launched a national political revolution toward negative campaigning, resulting in Sinclair's defeat. Prospects of losing control of the state to Democrats convinced agribusiness leaders to completely subjugate mobile workers. Throughout the 1930s, Republicans gave farmers more authority over their hired hands. Legislators eviscerated the Commission of Immigration and Housing. In 1933, the commission only had four agents to inspect 5,000 labor camps for almost 200,000 migrant workers.[100]

Conservative state leaders assessed what "Okies" had done to upset the race-caste system in the fields, and they took tough measures to bring American white migrants to heel. Determined to crush field unions and strikes, corporate and agricultural leaders created the Associated Farmers of California in 1934. The agribusiness juggernaut, which had very few farmers among its ranks, included railroads, paper producers, grocery stores, utilities, insurance and investment companies, banks, canneries, and sugar refineries. In 1935 the tide began to turn in the titanic battle between federal bureaucrats endeavoring to assist unemployed migrants and agriculturalists seeking to disempower them. State officials convinced federal bureaucrats to close cases and dispatch workers to farm counties at their own expense. Next, state officials successfully pressured the government to eliminate the Federal Transient Service, an agency that fed 77,000 indigent migrants a month.[101]

With the state now in charge of relief, officials created the State Emergency Relief Administration (SERA). Governor Merriam appointed farm leader Harold Pomeroy to head the new agency. Pomeroy slashed public aid to force recipients into the fields. In one case, a Los Angeles SERA caseworker discharged six hundred families from the relief rolls with instructions to drive 250 miles to the Imperial Valley and find work. SERA ensured the white migrants were sufficiently cowed when it ordered a thousand Filipinos to the desert to compete with them for jobs. State duplicity continued when Pomeroy ordered unemployment officers to cooperate with the Associated Farmers in all regards. With these steps, agribusiness firmly controlled farm laborers. Once the state handled relief, agribusiness launched an attack on SERA to undermine its authority and to transfer aid decisions to the county level. Agribusiness now had the best of all worlds. Only resident citizens were eligible for state and federal work relief projects. Local citizens got first shot at federal make-work positions and jobs on the All-American Canal. Although citizens and resident "aliens" were eligible for public aid, applicants had to meet the one-year residency requirement.[102]

By the summer of 1935, SERA began shifting public aid cases to the counties. The Imperial County SERA office acted as an employment agency, slashing its caseload and telling the unemployed to find work. When the local SERA director purged "aliens" unable to work from the caseload, his superiors transferred him and told the county it had to care for three hundred people incapable of working. SERA knew Imperial County had the state's lowest tax rate and could feed the destitute, especially since agribusiness received a windfall of federal construction projects. As SERA's funds ran short in October, the new county director purged Indians from the rolls and sent them to the Yuma reservation. He then vigorously slashed the rolls again and crowed he only had 450 cases, down from a high of 3,300 a year earlier. SERA transferred him for his excessive decisions.[103] Unrelenting criticism of direct federal relief bore fruit in early 1937 when federal aid ended for "transient unemployables"—four hundred families with sick parents, no parents, or women and children unable to work. With WPA jobs for resident men and women, the county transferred relief cases to charities, thus closing the chapter on Depression-era public assistance for the unemployed. By the end of the year, Imperial County had the lowest indigent expense in California, and it also ranked first in the lowest amount given to elderly indigents. Although agribusiness battled with liberals within SERA, it was generally happy with the agency's policies. Government programs had subsidized local taxpayers by feeding the unemployed and had given another subsidy to farmers when relief was denied during harvest season.[104]

New Deal programs reinforced the structural economic inequalities in California agricultural districts. Beyond helping migrants find jobs and dispensing groceries, governmental relief policies did little to benefit agricultural workers. Moreover, major legislation exempted agriculture from social and labor reforms. The farm lobby convinced legislators to exempt employers from the 1933 National Industrial Relations Act (outlawing child labor, setting a forty-hour week, and minimum wage), the 1935 Social Security Act (unemployment insurance, pensions, medical care for the poor), and the 1935 Wagner Act (recognizing unions and federal employer-labor mediation). Reformers were stuck—there was no counterbalance to the power of agribusiness. State officials and liberals failed to improve fieldworkers' lives.[105]

Governor Culbert Olson

California Republican hegemony ended in the election of 1938. Democrats swept the legislature, elected a US senator, and Culbert Olson became the first Democrat governor elected in forty-four years. Two decades passed before the next Democrat was elected governor. Olson campaigned on a New Deal program, openly courting the "Okie" vote. Although new white residents did not provide his margin of victory, they enabled Democrats to sweep the farm counties. During the contest, the archconservative California Citizens' Committee conducted a smear campaign similar to the one that helped defeat Upton Sinclair four years earlier. Liberals and white migrants were its chief targets. After the election, Imperial County supervisors endorsed a California Citizens' Committee report alleging white migrants threatened "the very foundation economically and socially, of California." In a letter to the *Brawley News*, a white migrant woman criticized the "carping attitude of the California Citizen's committee [who] is vicious and insulting to a large group of hard-working, self-respecting American families." Governor Olson spent his term in office sparring with agribusiness over relief implementation.[106]

Governor Olson appointed Carey McWilliams, the lawyer, liberal author, and archcritic of agribusiness, to head the Commission of Immigration and Housing. The year he took office, McWilliams published *Factories in the Fields: The Story of Migratory Farm Labor in California*, an indictment of corporate agriculture's concentration of power and exploitation of labor. McWilliams reversed the proagribusiness policies of his predecessor, offered a slew of reforms, and stepped up labor camp inspections. Agribusiness responded by labeling McWilliams "Agricultural Pest No. One, outranking pear blight and boll weevil." Equating McWilliams with agribusiness's top enemy—insects—had telling symbolism, for it linked nature and labor as commodities to be controlled. Liberals and conservatives fiercely criticized each other. John Steinbeck's *The Grapes of Wrath* (1939) immortalized the plight of poor whites, while the California Citizens' Committee pummeled Olson, white migrants, and liberals.[107]

Revealing the intensity of agribusiness intransigence against state assistance for fieldworkers, farm leaders vigorously opposed Governor Olson's program to feed malnourished children. In November 1939, SERA sent letters to 1,500 schools explaining a free hot lunch program for 200,000 undernourished children. SERA launched the program in cooperation with major federal agencies. A month later, the Imperial County's SERA director announced beginning in January the agency would hand out free hot lunches to 3,069 "undernourished and underprivileged children" in twenty-five schools. SERA pledged to pay 80 percent of the bill for extra help distributing the food. Brawley officials took the lead and rejected the food. When other schools in the county balked at the plan, the local SERA director failed to persuade Brawley leaders to at least give the food to children of fieldworkers.[108]

Conclusion

In the Imperial-Mexicali Valley, crops and racial labor went hand in hand. Over the course of forty years, farm leaders struggled to fashion the perfect workforce, with distinct jobs based on race. Settlers and absentee landlords depended on indigenous workers, itinerant whites, farm families, and a smattering of Mexican, European, South Asian, and Japanese migrant laborers. Increased acreage and the success of truck crops and cotton in the 1910s required seasonal harvesters, ushering in a shift in the workforce with the recruitment of Koreans, African Americans, Filipinos, and Puerto Ricans. South of the international boundary, the Colorado River Land Company recruited Chinese transnational laborers to grow and pick cotton. After World War I, farmers in the Imperial Valley, under the guidance of cotton interests in Arizona and the Colorado River Land Company, formed a cotton association to recruit temporary Mexican circular transnational migrants. Despite its failure, the cotton association was the first to successfully import and regulate labor and set wages. It also forecasted a model for consolidating power in the hands of large landholders, growers, and shippers. The cotton association's successful recruitment of braceros transformed the farm workforce with farmers depending primarily on Mexican migrants. Labor contractors also brought crews of Filipino and Japanese to compete for harvest jobs. Pitting workers against each other fostered racial antagonism, forestalled class unity, and kept wages down. Shippers segmented the labor market further by hiring white packers and demanding Japanese crop associations to hire white packers. The introduction of braceros in Mexicali instituted a critical mass to pressure politicians to enforce a law requiring 50 percent (later 80 percent) of all jobs be reserved for Mexicans, ending the era of Chinese transmigration.

Even though agribusiness in the United States received substantial assistance from state and federal agencies, government aid came with strings attached. State officials warned and occasionally fined farmers to send children to school and to maintain clean field camps. State officials also investigated wage and workplace complaints. Over time, agribusiness leaders decried meddlesome bureaucrats, who, with one hand, bestowed gifts, and with the other, insisted on improving the work and living conditions of field laborers. Agribusiness held the levers of power in the county, enabling them to rebuff demands to improve housing and sanitary conditions. Braceros in the Mexicali Valley succeeded in reducing Chinese labor competition, but during the Great Depression the Colorado River Land Company significantly cut back cotton production, resulting in widespread suffering and starvation. Unlike the United States, Mexico did not have the financial means to create a social safety net.

The Great Depression dramatically altered the calculus of control and power in the Imperial Valley when the exodus of Mexican repatriates coincided with the entry of white and black American migrants into California. Once economic hardship overwhelmed private charity, state and federal welfare programs alleviated the widespread suffering, but over time conservatives defeated proposals to raise the standard of living for fieldworkers. By the mid-1930s, California Republicans waged an all-out war against federal and state relief. Common hardship during the Great Depression and racial diversity among fieldworkers undermined the desert's caste-race hierarchy. "Okies" drained relief and took away jobs from residents and migrant workers. On the other hand, by demanding field jobs, white migrants undermined the racialized labor structure. Their poverty shocked the country and helped liberals push through public assistance for migratory fieldworkers. The New Deal social safety net, ironically, further subsidized corporate agriculture by feeding the unemployed, thereby redirecting responsibility away from employers and county taxpayers.

Fieldworkers and their families on both sides of the international line lived a life of poverty. When employers depressed wages it put tremendous pressure on men to provide for their families. Workers did not stand idly by while employers set wages, imported excess workers, instituted the exploitative labor contract system, and determined working and living conditions. Recognizing the cultural and economic decisions individuals and families made helps us understand why workers went on strike, risking jobs, deportation, and violence. As the next chapter demonstrates, agricultural workers on both sides of the border significantly shaped employer-worker relations.

CHAPTER 5

The Gospel of Labor Rebellion

> We have a strong, enthusiastic organization in the Valley, and the effectiveness of our work is evidenced by the lack of strikes in our county. We have an excellent formula for getting rid of cockroaches, grasshoppers, and C.I.O. agitators.
>
> Hugh Osborne, Imperial County Board of Supervisors, 1934

Agricultural workers in the Imperial-Mexicali Valley used a spectrum of resistance strategies to improve their wages and working conditions. Individually and collectively, workers' main source of power derived from giving or denying their labor to harvest the desert's agricultural bounty. Before the advent of organized farming associations, workers held the upper hand in determining wages. When farming organizations recruited migrants and reduced wages, workers responded by forming packing shed and field unions. Yet field and shed workers preferred negotiation and compromise rather than confrontation. Strikes by white packing shed laborers occurred frequently and often ended quickly and peacefully. When shed owners refused to negotiate with their employees, workers walked off the job and appealed for racial solidarity from their taxpaying and voting neighbors. As the nobility of agricultural labor, white shed workers demanded collective bargaining, and shippers, the police, the courts, and the press granted them respect and seriously considered their concerns.[1] Strikes and picketing by shed packers were tolerated, but Mexican field strikes challenged "law and order," and county officials brought the full weight of state and vigilante coercion and violence to crush them. Interracial solidarity and working-class collaboration floundered on the shoals of ethnic and racial animosity, nationalistic fervor, and cutthroat competition for jobs.

The Imperial Valley was a corridor of radicalism. Exploited workers, political radicals, labor organizers, social dissenters, and veterans of the Mexican Revolution entered from the four cardinal directions.[2] Disgruntled laborers devised ingenious methods to demonstrate their dissatisfaction. They stuffed cotton-picking sacks with dirt and rocks and pillaged produce from the fields. Laborers threatened and assaulted labor contractors and farmers, deserted abusive employers, filed wage complaints, petitioned the US and Mexican governments for redress, and denounced Mexican consuls who sought to defuse labor opposition. Spontaneous shed and field "wildcat" harvest disruptions were common tactics to gain concessions, express collective indignation, and demonstrate working-class power. Yet these practices were the last resort because delaying harvesting fast ripening crops brought devastating repercussions.

Worker-employer relations went through four periods in which both sides devised new strategies to meet changing circumstances. In the first period, labor scarcity led to competition for workers and high wages. Conflicts arose between bosses and individual workers or field crews. During this period, the Industrial Workers of the World (IWW) made the first attempt to organize fieldworkers in the Imperial Valley. The IWW's battle with desert agribusiness established a pattern of unionization and suppression that played out over the next two decades. Agribusiness considered it natural to organize for self-improvement, yet when fieldworkers joined a union and respectfully requested to negotiate wages and working conditions they were labeled "radicals." After World War I, the shift to Mexicans in the fields and whites in the packing sheds ushered in a second period of employer-worker relations. Plentiful, unorganized, migrant *braceros* (laborers) gave farmers the upper hand. Mexicans responded in 1922 by forming the valley's first Mexican union and launching on both sides of the international border a gigantic cantaloupe and cotton strike for higher wages. This was the first significant walkout in California by a Mexican union. In response, employers reigned in wages and controlled migrants with labor contractors.

The third stage in labor-employee conflict occurred during enormous field and shed disputes in 1928, 1930, 1933, and 1934. In 1928, Mexicans formed a union and went on strike, demanding the elimination of the labor contractor system, higher wages, and other workplace reforms. The walkout and the resultant crackdown established the template for future conflicts. In 1930, Mexicans and Filipinos organized the valley's first interracial union and refused to work in a bid to halt a slide in wages. Communist organizers and the American Civil Liberties Union entered the scene, beginning a five-year confrontation to reconfigure the balance of power in the fields. During the Great Depression, American and foreign-born migrants organized interracial unions and conducted enormous picketing campaigns.[3] Unable to match the power of interracial unity, corporate agriculture coalesced under the banner of the Associated Farmers. Farm leaders also appealed to the Mexican consul, Joaquín Terrazas, to create a Mexican union under his control as a wedge to divide the workers. At this point, representatives of both the US and Mexican governments joined with employers to suppress the labor revolt. Dispirited braceros crossed the border and demanded farmland, which launched the fourth era of labor rebellion in the Imperial-Mexicali Valley. For the rest of the decade produce shippers in the Imperial Valley disempowered white packing shed workers.

Industrial Workers of the World

Before the formation of the Industrial Workers of the World (IWW) in 1905, no sustained agricultural unionization had taken place in America. The IWW's egalitarian philosophy and critique of capitalist exploitation of the working class made it arguably the most radical union in American history. Its revolutionary vision to unite workers in one union regardless of ethnicity, race, and nationality drew a wide variety of discontented workers and ideological extremists. It organized mining, timber, industrial, and agricultural unions. IWW members, called "Wobblies," engaged in passive and spontaneous strikes and boycotts. The first IWW organizers came to the Imperial Valley in 1905 but failed to gain a foothold until 1909, when the union established four union halls. In that year, organizers launched a free speech campaign in Spokane and Fresno. Both cities passed ordinances restricting union activities and jailed organizers. After a protracted contest, the Wobblies won the right to establish union shops.[4] Working-class discontent in the American West paralleled revolutionary activity in Mexico. When the Mexican Revolution began in 1910, Imperial Valley Wobblies joined the magonista invasion of Baja California the following year.[5]

Over the next decade, the war of wills between the Wobblies and employers intensified. IWW organizers converged on San Diego to protest an ordinance restricting free speech. Mass arrests, beatings, and murders by vigilantes drove out the union organizers. After the San Diego debacle, the IWW continued organizing field hands. In 1913, the union drew national attention for participating in a strike in Wheatland, California, where a policeman, the district attorney, and two strikers fell dead. Despite a lack of evidence, the state convicted two Wobblies for the deaths. In the wake of the convictions, thousands of outraged workers joined the IWW. Farm counties punched the Wobblies with vagrancy and disturbing the peace ordinances.[6] In 1914, when the IWW stepped up its effort to organize an agricultural union in Imperial County, Calexico police arrested seven organizers for speaking in public and banished them from the county. Calexico passed an ordinance restricting street speaking without a permit. In the town of Imperial, Wobbly sympathizers brawled with agribusiness supporters after twenty-three unionists were arrested for disturbing the peace. When officials in Brawley permitted public speeches, Wobblies praised the town's "enlightenment" attitudes. However, the IWW struggled to build a countywide union. The following year, a crowd shouted down an organizer in Brawley. After a heated exchange he was allowed to state his "grievance." In El Centro, farmers and workers tussled when police arrested a Wobbly recruiter.[7]

Wobblies attempted to build a bread-and-butter union with the Agricultural Workers' Organization in 1915. The union made a valiant effort to establish a locale in the desert. A newspaper story, for instance, called Brawley's eastside the "wobbly" district. Union operatives in that town posted billboards in Spanish complaining of low wages and the high cost of living. Newspapers discredited the Industrial Workers of the World by lumping its members with criminals.[8] Outside the desert, the IWW recruited Filipinos, Mexicans, whites, and blacks and won major agricultural strikes in California. With America's entry into World War I in 1917, the government squashed the union's leadership for violating the Espionage and Selective Service Acts. A red scare led to twenty-three states passing criminal syndicalism laws. California's 1919 syndicalism law made it a felony for advocating or planning the violent overthrow of capitalism or the government. Therefore, belonging to the Industrial Workers of the World violated the syndicalism law and resulted in the imprisonment of 128 Wobblies after the war.[9] Although state repression mortally wounded the IWW, its members continued to test the power of desert farmers. In 1918, threats forced an organizer in El Centro to flee the town. The next year, Calipatria police arrested a man for promoting a labor meeting. A mob whisked him from jail, and after inflicting a savage beating, warned the unlucky fellow that "reds" and other troublemakers were not welcome. A rival union organizer expressed outrage because vigilante mistreatment of the IWW organizer threatened all unions. Six months later, after claiming residents "are ninety-nine and ten-tenths American," and radical "propaganda is not wanted here," a justice of the peace instructed "I. W. Blake" to go to jail or leave town by sunset.[10]

In the early 1920s, the IWW continued to shower the valley with organizers. At the start of the melon harvest, in 1921, Brawley police arrested Roscoe Thurman, the secretary of the Industrial Workers of the World. He "was held to answer on the charge of violating provisions of the espionage law" and distributing "Bolsheviki" literature to Mexican cantaloupe workers.[11] After a guilty verdict, Thurman appealed to the Supreme Court and won. Stung but defiant, the district attorney filed new charges, and for two years, he put his staff on the case without success.[12] In contrast to the IWW, agribusiness tolerated conservative or moderate unions as long as they excluded fieldworkers. During the war, three unions represented fourteen trades. Shed unionists hewed closely to the antiradical political winds. In a dispute between two shed unions, for example, a union organizer alleged a rival union harbored radicals and it better be careful, for "Federal agents are now in the valley investigating the I.W.W. literature being circulated."[13]

1922 Transnational Strike

The Wobblies were one of many radical groups passing throughout the borderlands preaching a gospel of rebellion to exploited laborers. Among the potent mix of assorted labor dissenters were grizzled veterans of the Mexican Revolution, political dissidents and labor organizers.[14] The rising tide of discontent against the foreign landowners and Asian laborers that caused the military garrison at Algodones to rebel in 1919 frightened the Colorado River Land Company. Over the next two years, militants challenged the CRLC's claim to landownership as a violation of the 1917 Constitution, drawing the ire of the company's general manager: "There is quite a lot of regular Bolshiviki's in and around Mexicali that are getting quite strong and they are claiming these lots and profess that they will take any lots they want" for themselves.[15] The company's stockholders, led by Harry Chandler, were stridently antiworker and antiunion. The CRLC private security force had long tussled with squatters and timber and cattle rustlers.[16]

The CRLC acted within a broader borderland security context. During the Mexican Revolution civil and military leaders of the northern district of Baja California hunted revolutionaries and arms smugglers along the border. The United States heightened tensions in the desert by stationing troops in Calexico. Governor Esteban Cantú countered American threats to annex the northern district by beefing up his well-armed cavalry and *campesino* (rural folk) militia. Cantú feared America's territorial designs. In the summer of 1920, when newspapers reported General Abelardo Rodríguez headed a Yaqui army on its way to oust Governor Cantú, Imperial Valley farmers pleaded with the US government to protect the Alamo Canal. Revealing disrespect for Mexico's sovereignty, the State and War Departments assured colonizers the government had contingency plans to seize the waterworks "from bands of marauders" who may try to destroy the irrigation system. Cantú ended the crisis when he resigned and went into exile in California. From his base in California, Esteban Cantú tried to retake the Mexicali Valley. In November 1921, for instance, after a week of skirmishing with insurrectionists, Abelardo Rodríguez, the district's military chief, executed the rebels.[17]

The following year, troops in Mexicali struggled to defeat a gigantic field strike that spilled over into the Imperial Valley. Although it may have been the first transnational labor walkout on the US-Mexico borderland, it has not yet garnered scholarly interest. Social unrest began in mid-1922, when revolutionary war veterans and repatriates seized foreign-owned farms. General Rodríguez met with the leader of the squatters, a former colonel in the army of Francisco "Pancho" Villa, and suggested his group should petition for land under the agrarian law. The colonel replied the squatters were not lawbreakers because foreigners could not own property near the border, and the agrarian agency in Tijuana had ignored their land petitions.[18]

Simultaneously, braceros secretly organized on both sides of the border. The importation of Mexicans after World War I resulted in a critical mass of landless, destitute field hands. Joe Molino related how Mexicans exploited white hatred of Japanese to gain higher wages. "In 1922[,] Garza, Ulloa and I organized a small union" in the Imperial Valley. "Most of the growers then were Japanese and the Americans did not care for them."[19] The union maximized the strike's potential by walking out during the simultaneous Mexicali cotton-picking harvest and Imperial's cantaloupe harvest.[20] Picketing began when cotton growers in Mexicali cut wages in half.[21] After Imperial Valley melon growers cut wages, the bracero union insisted on a slight pay increase. White shed packers walked off the job when employers insisted on reducing wages. At the time, Imperial Valley's triracial labor force numbered about one thousand Japanese, Mexican, and white "fruit tramps" who picked and packed melons. The Brawley Japanese Association hosted industry leaders to discuss the crisis. Employers decided to call for two thousand Japanese and Filipino replacement workers.[22]

Employers devised a six-pronged plan to end the labor stoppage. Local and outside Japanese filled open slots, while some employers raised wages slightly. The *Calexico Chronicle*, a merchant newspaper and mouthpiece for the antiunion Colorado River Land Company, wrote: "In Lower California the situation is declared to be so tense that the Mexican government has detailed soldiers to guard many of the ranches while laborers from the gulf, consisting of Chinese and Filipinos are being rushed by the ship load to take the place of the strikers." The final two strategies consisted of ordering troops to stop picketing in Mexicali, while Imperial Valley agribusiness red-baited the strikers: "what is declared to be one of the most tense labor situations in this section of the country has been uncovered and laid at the door of the Bolshevik and Red agitators" from Mexico. In Mexicali, picketers held out for a month before soldiers took control. Calexico police detained labor militants who fled martial law in Mexicali.[23] The harvest disruption raised wages in the Imperial Valley, but deterred field organizing for six years.[24]

Industrial farmers north of the border gleaned an important lesson from Mexico's repressive measures. Henceforth, Imperial Valley farming interests called upon the police and vigilantes to brutalize uncooperative workers. In the subsequent three years, authorities on both sides of the line banished critics from the region.[25] A few months after the 1922 Imperial-Mexicali Valley work outage, for example, Calexico private detectives caught "W. J. Willard, a proud Wobblie" with a membership card to prove it. A judge dispatched him to the chain gang. Two years later, spies informed police of two Wobblies passing though the valley. Police arrested the alleged radicals and charged them with criminal syndicalism. The arrest of H. Martínez for vagrancy in 1925 marked the end of an era for Wobbly involvement in the desert. When he told a Calexico judge he was a "proud" Wobbly, the jurist replied, "Well, well, well you must be a guest of the city for a month" on the chain gang.[26]

1928 Strike

The years between the 1922 strike and the next large-scale harvest protest in 1928 witnessed many changes in the labor market. The expansion of profitable truck crops translated into a shift to labor contractors who brought in crews of Mexicans and Filipinos. Labor contractors colluded with farmers to skim a portion of workers' pay and frequently absconded with wages. The contractor labor system precipitated the next major labor conflict in the Imperial Valley. The 1928 strike originated, however, from the Mexican government's desire to control its expatriate population in California by forming unions under the influence of Mexican consuls.[27] After the Mexican Revolution, the nation's new leaders brought trade unions under the government's control to foster a cooperative approach between employers and workers and to forestall union activity detrimental to social stability. Postrevolutionary leaders supported capitalist development and opposed socialist and communist inroads among the working class.[28]

Unlike American workers, braceros in Mexico had rights and protections enforced by courts and the state. In a recent study, William Suarez-Potts demonstrates how Mexico's workers skillfully used the 1917 Constitution's guarantee of working rights to create unions, demand negotiations with employers, go on strike, petition the government, and file lawsuits against employers for violating laws or agreements. Suarez-Potts traces the development of legal decisions supporting workers' rights and the rise of local arbitration boards to mediate disputes. By the late 1920s, Mexico established state and territorial arbitration boards to resolve work disruptions and enforce agreements.[29] This background on labor rights in Mexico sets the context for the actions of braceros in the Imperial-Mexicali Valley for the rest of this chapter.

In November 1927, the Federation of Mexican Societies of Southern California in Los Angeles passed a resolution to form Mexican trade unions. The following March, labor groups formed La Confederación de Uniones Obreros Mexicanos (Federation of Mexican Workers Unions or CUOM), the first major Mexican union in California representing twenty-two locals and three thousand workers. Although CUOM soon collapsed, Mexican organizing shocked agribusiness leaders who looked "upon Mexican laborers as tractable and docile persons."[30] Flush with optimism from the Mexican government's support for organized labor in the United States, the Mexican mutual aid society in El Centro formed La Unión de Trabajadores del Valle Imperial (the Imperial Valley Workers Union). In line with his government's directive, Mexican consul Carlos Ariza, stationed in Calexico, encouraged the organizing union to resist the contract labor system and to improve working conditions.[31] In April, Filemon González, president of El Centro's mutual aid society, presented a plan for a union. After the group expressed its support, González sent a letter to Consul Ariza explaining his motivations: "As a day laborer" and "a deep-rooted unionist I am well aware of the urgent need of some organization working for our improvement."[32]

With the melon harvest six weeks away, Mexican organizers publicized a meeting to sign up members. In a devastating blow to the union, the government abruptly discharged Consul Ariza when a widow complained he defrauded her of $500. The well-liked lawyer took a position with an El Centro law firm to recruit bracero clients, including the union.[33] Consul Ariza's replacement, acting vice consul Hermolao Torres, was present at the union's first meeting and in a foul mood because he was not informed of the organizing effort. Torres, who detested former Consul Ariza, became furious when workers hired Ariza as their lawyer. Filemon González, the union's secretary, tried to assuage the acting consul by sending him a letter explaining that the union was formed to protect the "Mexican colony" from exploitation. Recognizing the Mexican government's fear of radicalism, the union secretary promised, "The tendencies of this Union preclude ANY COMMUNISTIC IDEA." The conciliatory letter ended: "Please let us know whether our plan meets with your approval and let us have your instructions." The letter forecast a frosty relationship between the consul and the union.[34] Despite Consul Torres's displeasure, the union continued to organize. A circular read: "Year for year, there arise in work of all kind, for the toiler, innumerable difficulties preventing the latter from obtaining—except in rare cases—what is duly coming to him and what he claims by right and justice." Lawyers drew up the group's articles of incorporation and it signed up three thousand men, including merchants and labor contractors.[35]

The Mexican unionists acted in full accordance with how the conservative unions in the United States operated. In May, Filemon González sent polite "written demands" and a statement to farm leaders explaining why the union wanted a contract covering the cantaloupe, watermelon, and tomato harvests. "During the year we scarcely work 185 days, of which we acquire the sum of $555.00," which did not cover living expenses. Laborers requested the elimination of labor contractors, a pay raise, and insisted on weekly paydays, workers' medical compensation, and improved camp housing. Growers, however, did not respect the etiquette of collective bargaining with a field union. Instead, agribusiness instructed the police to eliminate the union.[36] When farmers began harassing workers, the union ordered its rank-and-file not to retaliate. González went to Mexicali and sent a telegram on May 5 to the Mexican president begging for protection. He said the telegram could not be sent from American soil "because we are not allowed to make certain statements contained in the present communication." He complained how Americans disregarded "every right due to us, crushing and insulting the laborers," provoking violence to give authorities the pretext to wreck the union. Americans "continue to arrest and mistreat, in various ways, the families of the workers. With weapons in their hands, [employers] try to force us to perform work." González informed the president: "The only accusation against us is: That we try to overthrow the order of things. And this is not true."[37] Actually, the growers were correct, the union did intend to overthrow the order of things by insisting on collective bargaining.

The unionists, however, received no response from the Mexican government, whose antifield strike policy in the Mexicali Valley mirrored its position across the international line. When dozens of workers refused to work, sheriffs evicted them from labor camps and arrested them for vagrancy. When picketers pled not guilty, the Brawley justice of the peace held them on $1,000 bail. Angry workers met in the streets and in pool halls "discussing loudly and vociferously the affairs of their union, their difficulties with their employers, and the jailing of their countrymen," wrote an investigator. Sheriff Charles Gillett deputized fifty shotgun-toting managers, foremen, and crop inspectors.[38] After mass jailings failed, officials unleashed a reign of terror. District Attorney Elmer Heald announced: "I don't care what price is demanded, but we can't stand by and let some gang go to a ranch and raise roughhouse." Heald handed recalcitrant workers to the Border Patrol for deportation on the grounds they were vagrants and liable to be a public charge. In order to reestablish control of its rank and file, the union's leaders called for a mass meeting at Brawley's Hidalgo Hall. On the day of the meeting, Sheriff Gillett jailed thirty men in El Centro. When informed of the Brawley rally, he promised to be "the chairman of that meeting." Police posted "operatives" in the east side of town where Mexicans lived to halt speeches or "disorder." Employers responded to strike leaflets by posting anonymous flyers written in Spanish warning workers that growers were prepared to ship in replacement workers. The union distributed placards in English and Spanish warning workers to beware of traitors and "knaves."[39] Meanwhile, the petulant vice consul Hermolao Torres sent Filemon González a second letter reiterating his feelings about being excluded from the union's formation and venting at former Consul Ariza's role in the union's founding.[40]

Farmers and union leaders expressed differing reasons for the conflict. A union leader told a reporter workers were eager to return to work. He blamed labor contractors, saying the union wanted to get rid of the middleman who was "getting fat without effort." A grower disagreed, placing the blame on the time-honored scapegoat of outside troublemakers. He reported "these agitators, or communists" had intimidated his workers into not working. In reply to accusations of radicalism, the union sent a letter to a newspaper: "we are not a bunch of bolsheviks or I. W. W.'s." The union simply wanted higher wages.[41] Sheriff Gillett arrested every Mexican he found on the street. When District Attorney Heald asked to meet with union leaders, the sheriff put them in jail. Furious, Heald berated the lawman and ordered the unionists released. Judges who did not agree with the mass jailings released laborers on their own recognizance. The sheriff tightened the noose by forbidding picketing, speeches, or passing out literature.[42] Gillett closed one "Negro" and three Mexican pool halls in Westmorland and padlocked the Brawley union office, declaring it illegal for "foreigners" to congregate on the streets or in public places. "There will be no more meetings," the sheriff intoned, "where there is likelihood of inciting riot." Equating braceros with children, he remarked: "The Mexicans are excitable and if idle will gather in groups to their own detriment as well as hindrance to work in the fields."[43] When Gillett saw a sign in a business: "Forty-eight Mexicans in jail—What for[?]—Nothing," he arrested a reporter for Mexico City and Los Angeles newspapers.[44]

During the police onslaught, Filemon González, a man angered by his government's abandonment, resorted to public sympathy, a tactic widely used in Mexico to garner support. He mailed letters to Mexican newspapers describing the situation. González wrote to *El Universal*, a prolabor periodical, we "have been threatened by armed authority with a view to compel them forcibly to work for as little as a single cent, if this be the pay. If we are dissatisfied, we should leave the country." Livid, he vented, "We desire that such oppression should not pass unnoticed. We never deserved, for one moment, those unnecessary threats." Americans, he claimed, blamed braceros for driving down wages, yet when they now organized "to improve our condition, we should not be accused of overthrowing the existing order." He ended his letter with a defiant statement to Mexico's leaders to help their expatriates: "We have certain rights to claim and we demand them."[45] González also wrote to a Mexican reporter in Calexico: "We are victims of endless injustice" and the Mexican government refused to help its citizens abroad. "Those in command tell us to retire from such meaningless proceedings and they threaten to have us imprisoned if we do not obey." González prayed: "As Mexican citizens, we do not doubt that our Government will intervene in this delicate matter which faces us." Hoping to avoid bloodshed and mass deportations, González told the union's members, "If they threaten us, bow your heads before the authorities." Mexicali's *El Nuevo Mundo* sharply criticized vigilantism and police brutality and warned: "the fruit harvest is approaching and [there is] only one type of laborer capable to resist the implacable heat of the sun—the MEXICAN laborer."[46]

Once the strike collapsed, Consul Torres took his American lawyer to win the release of workers. The judges lowered bail for all workers except for the union's Mexican American interpreter who advised workers on labor rights. After a week of rough treatment, the interpreter pled guilty to disturbing the peace, received a six-month suspended sentence, and fled the state. Union attorneys recommended guilty pleas in exchange for suspended sentences. When Consul Torres informed his superiors of the high bail and arbitrary arrests, Mexico protested to the US State Department, which called for an investigation. State officials sent statistician Dr. Louis Bloch to file a report.[47] After interviewing all parties, the investigator concluded that the union was not run by "reds." Bloch discovered that farmers had raised wages and he believed if the union survived it would "enlist" all the braceros in the county. He warned of future problems if the labor contract system, inadequate housing, and nonenforcement of the workers' compensation law were not addressed. Even the district attorney admitted the elimination of labor contractors was necessary. Bloch skewered growers: "Working conditions which give rise to real and justifiable grievances cannot be remedied by rough threats of deportation or by smooth talk of loyalty to the employers in exchange for past favors." He recommended addressing grievances instead of "resorting to the easier expedient of hurling time-worn and ineffective accusations of radicalism and red propagandism."[48]

Paul Taylor, the University of California, Berkeley, agricultural economist and advisor to the state Department of Industrial Relations, also interviewed the strike's main participants. Taylor spoke with Joe Molino, one of three organizers of the 1922 Mexican union mentioned previously. Molino said: "The Mexicans in the recent strike wanted better wages. There were no Bolsheviks and no red flags used in the parade." William López also insisted, "There were no reds and no talk of violence." Juan Estrada said the union wanted higher wages and wanted to eliminate the labor contractor. A union lawyer remarked, "There was no rebellion," laborers just wanted to negotiate with farmers.[49] Prominent whites expounded the agricorporate perspective. "The Mexicans wanted higher wages than could be paid and wanted to run this country," remarked the assistant district attorney. "It was Bolshevics and Communists from Mexico and Russia." A businessman added, "three men from Mexico stayed here as organizers whom we deported." District Attorney Heald said, "Why there was no rebellion or anything like that," but radicals lurked among them. He believed they had the right to meet, "but I don't know what three thousand Mexicans might do when they get together." A farm leader added, "Yes, the Mexicans won the strike demands. I believe either in settling a strike by conceding at once if you don't think you can win or else using any and every means to break it if you think you can." He believed in tough measures. "Carry big shot guns, not pistols. The trouble comes when [we] threaten and then don't do anything."[50]

The harvest labor confrontation had multiple outcomes. The deportation of union leaders rid the county of some militants and cowed others into silence. Although workers received a raise, many unionists were roughed up and spent time in jail. The indifference of Consul Torres undermined the union. He informed his superiors of the events on the ground and stood in court to ensure "fair" trials, but did nothing else. The Mexican government used consuls for multiple objectives, some of which worked at cross-purposes. Consuls were directed to create social welfare associations for mutual aid and to gain influence over expatriates. In the late 1920s consuls received orders to help establish unions as a way to deflate militancy. However, consuls insisted on resolving labor problems peacefully and swiftly to ensure a steady flow of dollars to supplement Mexico's economy. In the 1920s, Mexicans sent $58 million in remittances.[51]

White shed workers and Japanese shippers remained on the sideline during the strike. The Japanese and Mexicans despised each other and Japanese had to remain in good graces with whites in the aftermath of the Alien Land Law. White packers were not interested in interracial unity with Mexican fieldworkers because many were racists and undoubtedly feared packing shed owners who would undermine the high status of white shed workers by hiring Mexicans. Thus Japanese shippers, white shed workers, and farmers reinforced one another's position in the agricultural system. Another outcome was negative publicity. A local newspaper reported "the so-called labor strike" had "attracted country-wide attention," leading to the first-ever state investigation into a workplace disturbance. After state statistician Louis Bloch derided law enforcement's strong-arm tactics, the chastened district attorney dropped the remaining charges. The district attorney could act magnanimously, the harvest was almost over, and he did not want to embarrass the incoming consul, who urged his countrymen at a huge meeting at Miguel Hidalgo Hall in Brawley to obey US laws for they were guests in the country.[52]

One of the two final outcomes was a victory for the bracero union. Louis Bloch recommended ending the illegal 25 percent deductions from wages until the end of the harvest. Later that year Bloch told farmers the state would not tolerate cheating workers out of wages. He advised farmers to draw up a standardized contract spelling out terms of labor and pay. After a farm committee presented the state with a proposed contract, state officials announced the abuses of the contract labor system were resolved. The following year, the legislature passed a law requiring labor contractors to register and pay a $200 licensing fee. Eighteen months later, the number of labor complaints against contractors plunged. Finally, the pattern of labor militancy and agribusiness repression established a blueprint for how both sides would respond over the next decade. A few months after the strike, Chester Moore, the former head of the Imperial Valley Associated Labor Bureau, testified before a congressional hearing opposing legislation to place Mexico on the immigration quota list. His presentation helped scuttle the bill. Soon afterward, the Western Growers Association, the powerful Southern California farm organization, hired Moore as its executive secretary. During the Great Depression, Moore played a prominent role in opposing labor activism.[53]

1930 Strikes

With the 1928 Mexico quota bill defeated, farmers announced a labor shortage and requested the government waive visa and head tax fees for Mexican seasonal workers. When the government refused, cotton growers advertised for American pickers and white migrants streamed in, glutting the labor market, further exacerbating racial tensions and driving down wages. Filipinos and Mexicans set aside their bitter competition and united against the white migrants. Both groups organized separately but coordinated their activities, agreeing on wages for various crops and tasks.[54] When Mexican consul Edmundo Aragón heard of the union's plan to meet with the Western Growers and Shippers Protective Association (Western Growers), he failed to prevent the meeting. After listening to leaders of the Mexican-Filipino alliance, the Western Growers refused to meet again, saying they would not pay field labor "white man's" wages. The Western Growers also rejected Consul Aragón's request to mediate. Laying the groundwork for a crackdown, the press smeared the unionists as "reds" bent on overthrowing the government. In January 1930, Sheriff Gillett made scores of arrests and shuttered meeting venues, while judges threatened unemployed field hands to plead guilty in exchange for probation. Farm leaders requested state militia to quell "communist activities" and uproot "agitators." Arrests and intimidation squashed the uprising. Frustrated workers blamed the consul for cooperating with the district attorney.[55]

An unsuccessful strike may have faded from memory as the nation teetered on economic collapse, but communists from the Trade Union Unity League (TUUL) read about the strike's demise in the *Los Angeles Times* and the union sent Francis Waldron, Harry Harvey, and Tsuji Horiuchi to revive the valley's cantaloupe strike. Although these men had no experience organizing fieldworkers, they tapped into a hotbed of dissatisfaction. The communist organizers and field hands formed the Agricultural Workers' Industrial League (AWIL). Years later, Harry Harvey recalled Mexicans and Filipinos "warned us against the stool pigeons, gave us information, and let us know the temperature of the workers in the various fields, just how hot or cold they were toward the strike." Harvey remembered: "No sacrifice, in the matter of getting them better living and working conditions, seemed too great to make." When the AWIL inquired about speaking at the defunct bracero union's hall, its horrified leaders refused and informed the consul of the presence of interlopers.[56]

The AWIL drew up a list of demands and requested a meeting with employers. After agribusiness leaders refused to negotiate, the union announced a work stoppage. The AWIL divided the Mexican community. The AWIL's members hustled speakers into the defunct union's hall against the wishes of more cautious workers. The walkout infuriated the consul and embarrassed conservative Mexicans, who refused to sell food or extend credit to the strikers. Hoping to avoid another state investigation, the press remained silent during the crackdown, until the three communists were arrested.[57]

Responding to pleas for federal assistance, Charles Connell of the Department of Labor's Conciliation Service devised a plan to undermine the AWIL, while working with officials and the consul to purge unruly workers. Connell called for a tough line from law enforcement, cooperation with the Mexican consul to repatriate radicals, and forging of links with the state-controlled union in Mexicali. Connell's work paid off quickly. A. Mejía, the president of the dormant bracero union, sent telegrams to the Mexican government requesting land. When Mejía received a positive response, an overflow crowd spilled onto the street cheering wildly. Mejía declared: "We believe that colonization on a Mexican agricultural project will be better than 30 to 35 cents an hour in Imperial valley." He shut the hall to all meetings except those dealing with repatriation. Police defeated the union when they intercepted truckloads of food and tents destined for unemployed workers.[58] Thus began the shift in circular Mexican migration on the California borderlands, as hundreds of expatriates voluntarily returned home. Unfortunately, when braceros arrived in Mexicali, the government ignored their petitions for land.[59] Adding to their misery, landowners in Mexicali reduced cotton production in the face of dire economic conditions. Landless workers, led by Felipa Velásquez viuda de Arellano, a socialist, formed an agrarian committee to petition for farms. Critics mocked the governor for his slavish support of the foreign landowners and his unwillingness to help the destitute. The governor silenced the troublemakers by sentencing twenty-six men, Felipa, and her eight children to the notorious penal colony Islas Marías.[60]

Meanwhile, at the height of the Imperial Valley lettuce harvest, El Centro packers refused to work until they received a slight raise. When the Agricultural Workers' Industrial League tried to persuade shed packers to join them, packers refused to associate with them. Former members of the defunct bracero union and some Filipinos saw an opportunity to forge an alliance with white shed workers by promising to support packers if they helped raise field wages. Packers agreed and nine hundred fieldworkers walked off the job in a sympathy strike. With packers and two rival groups of fieldworkers not working, it became increasingly difficult to shape public opinion. The leaders of the revived bracero union released a bulletin: "in no way is the strike being conducted by the red International of labor unions or any organization affiliated with the Agricultural Workers' Industrial League." The unprecedented unity of Mexican and Filipino fieldworkers and white shed workers galvanized shippers and a Department of Labor's conciliator to end the strike by pitting two rival packing shed unions against each other.[61]

After the lettuce harvest, the Trade Union Unity League sent a lawyer to represent jailed members of the union. Police harassment convinced the lawyer to ask the Reverend Clinton Taft, the head of the American Civil Liberties Union (ACLU) in Los Angeles, to free jailed workers. When Clinton Taft went to speak with the sheriff, the lawman assaulted him. Taft's associates voiced outrage, forcing the Imperial County district attorney to release the prisoners.[62] Police routinely assaulted prisoners and "undesirable" visitors. Pummeling Clinton Taft, however, infuriated liberal lawyers in Los Angeles, which shifted the power dynamics in the Imperial Valley. From this point forward, the ACLU cooperated with lawyers for left-leaning field unionists to resolve civil liberties violations and to promote field unionization.

The communist-affiliated AWIL began a union drive in March for the upcoming cantaloupe season. Organizers signed up Mexicans and Filipinos, and a trickle of whites, Japanese, and blacks attended strategizing sessions. Police had broad authority to break up meetings since the Brawley police chief defined disturbing the peace as a gathering of more than two persons. Police barged into workers' homes and arrested union members. Police closed public places where fieldworkers met and hired gunmen from Arizona and Texas who guarded sheds, bridges, public buildings, fields, and warehouses. The district attorney pitched in by hiring spies to infiltrate the AWIL. The strike collapsed in the face of the antiunion onslaught. The battle between the union and the farming autocracy moved to the courts, where the district attorney used union organizing literature to convince the jury the objective of the Communist Party was to overthrow capitalism and democracy, acts that violated the criminal syndicalism law. Unsurprisingly, the jury found the men guilty and they received the maximum sentence for each count (fourteen years per charge), which the ACLU considered the gravest violations of civil liberties in America that year.[63]

State-sanctioned violence and civil liberties violations had foiled another bid to improve the lives of field workers. District Attorney Heald won the first convictions of communist organizers for criminal syndicalism. The commander of a local American Legion post addressed a Legion convention: "The way to kill the red plague is to dynamite it out. That's what we did in Imperial County," he bragged. "The judge who tried the Communists was a legionnaire, 50 per cent of the jurors were war veterans. What chance did the Communists have? That's the way we stamped it out in our county." When AWIL organizers returned later that summer, Mexican consul Aragón established conservative organizations to marginalize Mexican radicals. The convictions

Agricultural Workers' Industrial League defendants outside county courthouse, 1930. Courtesy H15–02_J2–2, Imperial County Historical Society, Pioneers Museum, Imperial, California.

of the AWIL union leaders reinforced the belief within shed unions that affiliation with fieldworkers was the kiss of death.[64]

Great Strikes of 1934

California lay at the center of most of the nation's agricultural strikes during the Great Depression. From January 1933 to the summer of 1939, there were 180 labor disputes in thirty-four of the state's fifty-eight counties, involving a third of California's farmworkers. After the collapse of the AWIL, labor activists established the Cannery and Agricultural Workers Industrial Union (CAWIU). It participated in or led most of the harvest walkouts, leaving a labor record unparalleled in the state's history. Passage of the National Industrial Recovery Act (NIRA) in June 1933 sanctioned collective bargaining between industrial workers and employers. Reports of NIRA's hour and minimum wage provisions failing to cover agriculture resulted in thirty-seven strikes in California, with the CAWIU playing a key role in most of them. The largest work stoppage occurred in the San Joaquin Valley cotton fields in September and October 1933. The CAWIU's 18,000 Mexican, Filipino, black, and white members organized a picketing zone stretching over a hundred miles and spanning four counties. Federal mediators brokered a slight wage increase after vigilantes murdered three picketers.[65] The San Joaquin Valley cotton harvest uprising inspired Imperial Valley's braceros to push for wage increases and workplace reforms. Workers of color kept abreast of the New Deal laws through newspapers and radio. When Mexicans in Brawley staged "public demonstrations" for the New Deal, Democrats welcomed their activism and white businessmen urged members of Club Demócrata to organize a union. Simultaneously, following a long history of co-opting workers, some farmers encouraged the revival of the disbanded bracero union in Brawley—but under farmer control—as a foil against the left-leaning CAWIU.[66]

Unbowed migrants created La Unión de Trabajadores del Valle Imperial (La Unión) in October 1933. Headquartered in Brawley, La Unión demanded a two-dollar-a-day wage and negotiations for a wide range of jobs. When growers refused to negotiate, the union flexed its muscles with a one-day strike. Faced once again with an obdurate foe, agribusiness shifted tactics and recruited the archconservative Mexican consul Joaquín Terrazas to discipline field hands. La Unión's formation infuriated Consul Terrazas, who accepted an agribusiness proposal to negotiate an agreement to eviscerate the union. Consul Terrazas and Alvin Jack, of the Western Growers and Shippers Protective Association (Western Growers), agreed to a $2 per day wage for nine hours of work.[67] Alvin Jack praised Consul Terrazas as reasonable "and just in the handling of the local situation and it was a pleasure for the growers to meet with him." Terrazas, a member of the wealthy conservative Terrazas clan in Mexico, was determined to stamp out radicalism and promote nationalistic sentiment among Mexican expatriates. He performed his official duties while profiting handsomely from his position. The American consul in Mexicali discovered that Terrazas ran a number of rackets, including charging a $20 "fee" to help his countrymen enter the country legally.[68] Laborers accepted the consul-brokered deal but grew angry when growers ignored the agreement. During the pea harvest, Mexicans attended a "monster" meeting at Brawley's Azteca Hall. Ignoring the consul, they voted to withhold their labor. Spurred on by newspaper editorials demanding tough measures, the police arrested picketers. The next day, Consul Terrazas brokered a 10 percent wage increase. Furious that field hands had ignored his agreement—and perhaps embarrassed by his inability to avert the walkout—Terrazas ordered strikers to apologize. Nevertheless, protests continued because farmers discharged braceros and hired American migrants at lower wages.[69]

Seeking to isolate the militants and seize control of La Unión, consul Joaquín Terrazas and his handpicked union representatives met with Alex Jack regarding farmers' failure to abide by the contract. The meeting ended with Jack admitting the Western Growers could not compel farmers to uphold the agreement. Fed up with the consul, dissidents traveled to Los Angeles and persuaded the Cannery and Agricultural Workers Industrial Union (CAWIU) to help them defeat the Terrazas-agribusiness alliance. The CAWIU sent two organizers who collaborated with braceros to establish a union affiliate. Next, the unionists rented Azteca Hall in Brawley for their headquarters.[70] In response to the threat from the left-leaning union, the consul's handpicked negotiators representing La Unión met with the Western Growers in early January 1934. Agribusiness rejected a request for 35 cents an hour and three days later when La Unión's membership heard the news they walked out en masse to join the CAWIU. The fed-up workers articulated their position: "We have been told that an increase of salary is not possible" because shippers were losing money. "We understand it perfectly," however, laborers could not depend on the ups and downs of the market to survive. The CAWIU demanded 35 cents an hour, transportation to the fields, free and clean water, a minimum five-hour day, equal pay for all workers regardless of age or sex, the elimination of labor contractors, and most importantly union recognition as the bargaining agent. The CAWIU lacked recognition anywhere and desert farmers refused to set a precedent. As soon as the union ordered its members off the job, thousands of field hands who caravanned to a rally in El Centro were attacked by police and vigilantes.[71]

The outcome of the labor impasse depended, in large part, on the decision of shed workers to support the communist-affiliated union. Brawley lettuce packinghouses took advantage of the unrest and walked out for higher wages, though they refused to cooperate with the CAWIU. Meanwhile, skilled workers in the county obtained a charter from the AFL to establish the Central Labor Council. The Regional Labor Board sent a representative of the AFL and a state labor commissioner to investigate. Upon their arrival, the head of the county Highway Patrol scolded them: "You men get out of here. You are hurting our work. We don't want conciliation. We know how to handle these people and where we find trouble makers we will drive them out if we have to 'sap' them." The Central Labor Board stopped its investigation after the Board of Supervisors complained to the governor about its meddling.[72]

Women took leadership roles organizing picketers, speaking at rallies, and browbeating scabs to quit working. A *Los Angeles Times* reporter commented: "The leaders are using women for their front line defenses. They know the police won't be too rough with them, and by the time the cops have waded through a cordon of amazons the real leaders have disappeared." Police and "special officers" from the coast rounded up 115 picketers at a shed and marched them six miles to the El Centro jail. Two women who led the "army" of picketers on the way to jail shouted pleasantries at passing motorists. When the procession hit town, consul Joaquín Terrazas implored them to return to work. Police released the picketers after they promised to ignore the Cannery and Agricultural Workers Industrial Union. Incapable of incarcerating so many people, police permitted picketing. The county supervisors called for state troops, but the governor demurred, sending instead the head of the state's Highway Patrol.[73]

Police decided to raid the union headquarters and arrest five union leaders. With warrants in hand, 150 police and vigilantes converged on Azteca Hall, breaking windows and firing tear gas into the building. Screaming occupants fled as police, vigilantes, and workers brawled. A baby and an adult died from tear gas inhalation. Authorities ransacked the hall and stole equipment, furniture, food, and supplies. "Before leaving the interior of the hall, more tear gas bombs and pestilent liquids were thrown on the stage and floor," staining it with foul odors. Later that day, police "threw tear gas bombs inside," and the fire department flooded the building. Across the eastside, police invaded homes and assaulted and arrested Mexicans. Police herded men and women loitering on the street into a stockade, arrested Mexican reporters, and "questioned severely" consulate investigators.[74]

Conservative Mexican leaders, a representative of the National Labor Board, and Consul Terrazas met with agriculturalists to suggest arbitration. Farm leaders refused, claiming there was no strike, and lawmen had the situation in hand. Agribusiness refused to acknowledge a strike out of fear of a state or federal intervention. After farm representatives praised Consul Terrazas for bringing about splendid "harmonious relations" between workers and employers, the consul agreed with the assessment that communist outsiders rather than exploited migrants were behind the uprising: "It is the agitators who have come into the valley to frighten my people that has caused this difficulty," asserted Terrazas. "I think when the valley authorities get rid of them the situation will be cleared."[75] Consul Terrazas did not operate on his own volition, for consuls were closely monitored and they implemented the policies of their government. Farm organs, like the *Brawley News*, warned "new agitators and communistic sympathizers" were on the way to reinforce the union. It warned "good citizens" not to help field unionists for they would "tear down government itself and carry on a reign of terror." The newspaper encouraged police and deputized residents to stamp out radicalism: "The 'red' flag of rebellion is not going to supplant the Stars and Stripes in Imperial valley." Taking its cue, two hundred police and hundreds of vigilantes patrolled the roads arresting picketers.[76]

Reports of the flagrant violation of civil liberties galvanized liberals in Los Angeles, who demanded a federal investigation. Stung by the criticism, federal officials sent Campbell McCullough, the secretary of the Regional Labor Board, to interview unemployed workers and tour field housing. In his report, McCullough described low wages, labor contractor abuses, and unsanitary camps. He exposed the interlocking structure of power among police, judges, and agribusiness. Most county officials were farm landowners. B. A. Harrigan, the agricultural commissioner, for instance, owned four farms leased to Japanese. Brawley justice of the peace H. B. Griffin leased his farm to Japanese. McCullough wrote: "It is alleged that no worker can hope to get justice" from

Judge Griffin because he considered "$1.00 a day is enough for a Mexican field worker." When questioned about worker poverty, B. A. Harrigan replied starvation was good for migrants. Disgusted, McCullough recommended a state-federal panel to recommend reforms.[77]

Soon after McCullough departed, police captured the top three Cannery and Agricultural Workers Industrial Union organizers, and the union called off the strike. The CAWIU's lawyers and the American Civil Liberties Union publicized police and vigilante violence and demanded civil liberties protection. A. L. Wirin, a nationally known ACLU lawyer, announced a meeting in Azteca Hall to advise workers of their constitutional right to organize. On the day of the meeting, the *Brawley News* published a warning of the impending communist invasion. Hundreds of vigilantes poured into town, while two thousand Mexicans and Filipinos waited in the hall for Wirin to show up.[78] Agriculturalists believed liberal lawyers posed a greater threat than farmworkers, for laborers could be handled roughly, while well-connected outside lawyers could influence public opinion. The first step in defeating the union, then, was vanquishing the ACLU.

Before the meeting in Azteca Hall, vigilantes kidnapped Wirin in a restaurant lobby and spirited him to the desert, where he was beaten, robbed, and abandoned. A passing motorist found Wirin and drove him to his hotel. Earlier in the day, vigilantes prevented "agitators" at the same hotel from attending the meeting. With drawn pistols, drunken vigilantes terrorized the liberal activists outside the hotel. One of the women activists was sexually assaulted with a handgun. Fearing for his safety, Wirin promised never to return to the desert in exchange for safe passage out of the county. A Highway Patrol officer and a caravan of vigilantes escorted him to San Diego. At Azteca Hall, a local minister and a representative of the ACLU informed the crowd that meetings and unions were legal, yet as braceros filed out police assaulted and arrested them. Wirin sent telegrams to the president and the governor demanding state and federal investigations into his rough treatment and kidnapping.[79]

A barrage of complaints from liberals jolted authorities to send a team from the federal Labor Relations Board in late January 1934. All three men of the Leonard Commission had solid liberal capitalist credentials. They devised a plan to destroy the communist threat by improving living and working conditions. Commissioners held public hearings and met privately with many people. The Leonard Commission's report savaged agriculturalists for sanctioning police brutality and violations of civil liberties. Commissioners made numerous recommendations, including the creation of a federal labor mediation board, a federal administrator to oversee employment, the protection of unions, collective bargaining, housing improvements, migrant education, subsistence farms for the unemployed, and a federal investigation into the Azteca Hall raid. The report illuminated the divergent positions of the combatants. Employers assured the commission that fieldworkers were "easily led," "docile," "contented," but influenced by "outside agitators." Filipino and Mexican leaders scoffed at such absurdities. Investigators noted Mexicans did "not favor any peonage attitude toward their race."[80] The report stunned the district attorney, who replied weakly: "The Communist party convicted itself in our own courtroom three years ago" during a strike. Communists criticized the report as well, saying the commission intended to break the union. To forestall federal intervention, farm leaders asked consul Joaquín Terrazas to organize Brawley's Mexican social and economic leaders to counter radicalism, resulting in the formation of La Comité Mexicano de Bienestar. Under the consul's direction, La Comité started writing a report to the secretary of labor explaining labor relations from the Mexican perspective.[81]

As the Leonard Commission and La Comité wrote reports in February 1934, three thousand Mexican, Filipino, white, and black pea pickers walked off the job. The head of the Federal Employment Bureau said conflict was inevitable because farmers let "a collection of poor white trash" stay on ranches over the winter working for thirty cents a day while supported by federal public assistance. The white American migrants subscribed to conservative, white supremacist beliefs when they arrived in California, but years of privation radicalized them. White fieldworker discontent reshuffled the power structure in the state's farm counties. The pea strike reveals white, and a smattering of black migrants, forged interracial working-class alliances to challenge agribusiness hegemony. Police were overwhelmed by an unprecedented show of force. Negative publicity of past abuses restrained vigilantism. One grower growled, "It isn't a strike, but a young Red revolution." As squadrons of picketers hurled threats and insults at those still working, relief agents distributed grocery tickets to 1,892 families.[82]

Within a week the picketing caravans numbered four hundred vehicles. Some shouted, "strike if you have American blood." Farmers were more terrified than during the lettuce walkout because unionists were "mostly white laborers, and the number of 'red' agitators is thought greater." Replacement workers, known as scabs, were in immediate danger. "They cut off my hand—maybe my head," said an old picker. He told growers: "You can take care of me here, but up north where I go, they get me." Another said, "They tell us all the time that wherever we go, they know us and they get us there. If we go into the fields, they see us. They know us. They carry field glasses." Picketers rejected proposals that did not recognize the CAWIU as their bargaining agent. The county supervisors waded into the fray passing an ordinance making it illegal to obstruct the roads, enter fields without the owner's permission, or to threaten workers.[83]

Evicted from labor camps, thousands of field hands squatted on federal land on the edge of the valley. It was a wise strategy to camp on government property to avoid arrest for vagrancy. A female reporter wrote a feature article on the strike camp. When she and three others drove in, a young woman wearing "dirty blue overalls and a man's grey workshirt" stood on a pickup addressing a crowd. She pleaded for those with money to contribute to a community fund to buy food. With "hair flying and arms waving wildly," she yelled, "We've got those farmers licked if we just hang together." She hollered, the farmers were organized, workers should be, too. "If they aren't how the devil do they whip us every time we strike?" When a man said something, she kicked him in the shoulder and kept right on haranguing the farmers. After she was done the sheriff explained the new antipicketing ordinance.[84]

A researcher toured nine squatter camps on government land estimating the racial composition at 70 percent Mexican, 25 percent white, and 5 percent Filipino and black. Employers heavily advertised for labor, and there were two or three pickers for every job. "They say this [strike] is run by Communists," said one informant, but "I haven't seen any around." Another worker intoned: "We guarded the camp last night. Only one car tried to get in and we swarmed on it with clubs, knives and guns." The researcher's handwritten notes mentioned one caravan of sixty-four cars extending for seven miles. Picketers shouted insults and threats to workers: "come out you scabs"; "you can't stay there forever and we'll get you when you come out," all the while shaking fists. One worker attested: "The Filipinos and Mexicans won't scab," while another added, "One field quit and they come down the road honking their horns," so we joined them. A disgruntled white migrant summarized the shift in the political landscape: "Roosevelt is behind us. He says go ahead and strike. I'll help you get out from under. He feeds us too. Don't let anyone tell you he isn't our friend."[85] When the antipicketing ordinance had no effect, the state health board declared the striker camps a health hazard. Deputies evicted the occupants and burned the shanties. Farmers refused to hire the evicted strikers and police arrested pickers. Local authorities gave food and gasoline to those who promised to leave. Farm leaders had defeated the second work stoppage in two months.[86]

Racial solidarity posed the greatest threat to agribusiness domination over workers. The potential of an interracial alliance led by communists frightened the Mexican and US governments as well. To smash the CAWIU, the National Labor Board and agribusiness leaders asked Consul Terrazas to form a Mexican union under his control, essentially inserting a functionary of the Mexican government into US labor relations. It was an illegal and unprecedented proposal, yet these were unprecedented times. After Terrazas discussed the proposal with his superiors, Luis Valdés of the Mexicali-based Labor Board of Baja California, an agency that frequently resolved labor crises, announced his support.[87] Consul Terrazas rushed to isolate militants by requiring a minimum of six months' residency in the county as a prerequisite to join his union.

Only a trickle of workers joined when word got out that the union's officers were farm foremen and labor contractors. Growers distributed union membership cards and fired and evicted workers from field camps who refused to sign up. Hatred of Consul Terrazas was so acute he held meetings outside of ethnic Mexican neighborhoods. A letter to the Mexican government complained of Terrazas cooperating with growers, "placing us at the mercy of the companies who will pay us what they please." The criticism fell on deaf ears as the government ordered the consul to proceed. Terrazas rented the Brawley school auditorium and told 1,500 listeners of his plan to revive the local bracero union to protect expatriates from communists. Reflecting the consul's authority in the community, the leaders of Mexican social and civic organizations sat on the auditorium platform. A representative of the Mexicali Labor Board remarked: "As soon as Mexican workers are organized in Brawley, the whole hearted moral and material support of 2,000,000 organized laborers of the Unions of Mexico will be behind them."[88]

While Terrazas built up his union, California agribusiness formed the Associated Farmers to eviscerate the Cannery and Agricultural Workers Industrial Union. Originating in Imperial County, it was the end product in the evolutionary process of agribusiness power in California. As perhaps the most powerful agricultural association in American history, it spread propaganda, criticized liberals and communists, organized vigilantes, bought weapons, lobbied against public aid for fieldworkers, and supported strike-busting ordinances. The Imperial Valley branch of the Associated Farmers created the Anti-Communism Association, a six-thousand-member unit, headed by Charles Nice, a former American Legion leader and the county indigent commissioner, where he supervised the chain gang. Nice had seven thousand Legionnaire vigilantes at his command because he volunteered as the county's veterans' social benefits officer. Farm leaders considered the CAWIU an enemy of the state, and they named a captain in every community to "have complete control over his particular unit" with authority to stamp out radicalism.[89]

White shed packers expressed alarm at the dual threat of Consul Terrazas's union and the Associated Farmers, recognizing the harmony between these groups threatened their tenuous position in the agricultural labor hierarchy. Yet conservative packers also feared communist inroads. In March 1934, the moderate American Federation of Labor's Central Labor Council held a spirited debate on whether to support the CAWIU. Strong yet "tactful" pleading kept it from passing a resolution in support of the radical union. After a week of pressure, the council took a stand against the CAWIU. The AFL rejected Consul Terrazas's foray into US labor relations, and it began organizing a field union of Mexican and black laborers. When the AFL asked Consul Terrazas if he would help them, he refused, remarking, "he knew his people and they'd object to the fees and direction" from whites. The AFL agricultural union recruited a Mexican organizer, but it could not compete with Terrazas's union. A federal labor official later disclosed how the consul helped break the CAWUI in Imperial County: "under the authority" of the Mexican government, Terrazas organized Mexican laborers even though the AFL protested Consul Terrazas's involvement in domestic affairs violated American law.[90]

Blindsided by the alliance between Consul Terrazas and industrial agriculture, the American Civil Liberties Union urged fieldworkers to organize under their own leadership. Shielded by an injunction protecting it from police and vigilante interference, a huge crowd listened to speeches at Azteca Hall. County officials and farm leaders in attendance saw every worker raise his hand when asked if they wanted to join the Cannery and Agricultural Workers Industrial Union. When the speakers left, vigilantes threatened to kill the speakers if they returned. The next day, farm leader Alvin Jack sounded the alarm. "Anyone present at the Azteca Hall, Saturday would be justified in thinking they were in some country other than America." Jack claimed the union planned to burn sheds and destroy bridges to disrupt the upcoming cantaloupe harvest. The CAWIU and the ACLU distributed flyers promoting a meeting in Calexico. The day before the meeting, police arrested a slew of civil liberties lawyers and CAWIU organizers throughout the county. Vigilantes beat and kidnapped lawyers representing the outside groups. One ACLU lawyer escaped vigilantes by taking sanctuary in the Calexico jail. The Anti-Communist Association packed the Calexico meeting in a show of force. The strike and violations of civil liberties drew national attention when ACLU lawyers published their mistreatment in the liberal magazine the *Nation*.[91]

With the ACLU lawyers and the CAWIU leadership on the run or in jail, a few days later, agribusiness negotiators met with the leadership of Consul Terrazas's union to formalize a cantaloupe contract incorporating many of the CAWIU's demands. In the agreement, which farmers did not have to abide by, employers promised to use Consul Terrazas's union exclusively. It set wages, eliminated labor contractors, and pledged to uphold sanitation laws in labor camps, among other items. A resolution from the agricultural leaders praised Terrazas "for his effort in aiding the preservation of industrial peace" and mailed a copy to the Mexican government. With the agreement finalized, Terrazas and farm leaders rushed to dismantle the CAWIU before the federal government announced its response to the Leonard Commission's report.[92]

New Dealers in Washington, DC, avoided antagonizing the powerful agribusiness lobby. Federal officials disagreed with the Leonard Commission's recommendations and hoped to use conciliation to bypass the communist-affiliated union (like it had done during the 1930 strike), but this time by suggesting labor reforms agreeable to farm interests. Federal officials appointed Pelham Glassford as the strike conciliator. Glassford, a Phoenix, Arizona, grower, retired general, and former police superintendent of the District of

Columbia, had a reputation negotiating intractable disputes. After failing to reverse Glassford's appointment, the Associated Farmers, the state Farm Bureau, and the state Chamber of Commerce appointed a three-man committee to conduct a parallel investigation.[93]

After meeting Glassford, county officials believed he would give them a "square" deal. Glassford and the profarmer committee held joint hearings. Members of the Cannery and Agricultural Workers Industrial Union spoke of low pay and horrible working and living conditions. They recounted vigilante terror and ridiculed Terrazas's union. Lawyers for the CAWIU, the American Civil Liberties Union, and the American Federation of Labor provided supporting testimony. When Agnes Bailey, a Mexican American representing the CAWIU, gave investigators the union's demands, Glassford accused communists of using workers for their political agenda. Growers pledged never to negotiate with the CAWIU. Glassford brushed aside the farmer-consul cantaloupe contract and announced he would establish a code concerning wages, working conditions, and camp housing.[94] In *Open Forum*, the ACLU's news outlet, Chester Williams warned: "This is the struggle against a ruthless American brand of Fascism. Fascism is best fought at specific points. Imperial Valley is one section where the fascist spirit and method have triumphed. It must be put down there with every bit of power we can muster." The agribusiness committee's report supported Chester Williams's analysis. It exonerated farmers for low wages and deplorable living conditions and justified violations of civil liberties. The report alleged the CAWIU sought to cripple agriculture as a first step in overthrowing the government and it recommended draconian measures to extinguish radicalism.[95]

Upon the request of the Board of Supervisors, federal conciliator Pelham Glassford arbitrated complaints prior to the cantaloupe harvest. Glassford relied on persuasion because his superiors refused to give him clear instructions or police power. Since Consul Terrazas's union was a farce and the growers and shippers association could not compel farmers to comply with the labor contract, Glassford took grievances to individual growers or to the labor commissioner. The CAWIU knew Glassford sought to undermine the union so it smeared him as a toadie of the agribusiness fraternity. Glassford retaliated with seven thousand flyers blasting the union's circulars as "full of falsehoods and erroneous quotations" and asserting the CAWIU had "no interest in the welfare of the workers. Their only objective is to create dissention, destroy private property and foment a strike." Glassford warned laborers not to "follow the advice of these vile agitators. To do so will bring destitution, despair and hunger, and will add nothing to the welfare of the workers."[96]

Liberal lawyers and the liberal union used Glassford's presence to urge workers to select which union they wanted to represent them. They set the election for May 5 (Cinco de Mayo), a Mexican holiday, but vigilantes threatened outside speakers and the election was canceled. Stunned by the oppressive measures, Glassford criticized the police and submitted a report chronicling police intimidation and assaults on ACLU lawyers. The conciliator was no longer considered an ally, and the Associated Farmers demanded his recall. The Cinco de Mayo fiasco and the Western Growers' inability to enforce the labor agreement turned Glassford against agribusiness. On May 7, he announced an election to allow workers to choose their own union. Complaints from farm leaders, however, forced him to call off the election. Glassford now changed his opinion of agribusiness and consul Joaquín Terrazas's union. An independent investigator scribbled notes from a conversation with Glassford: "the Mexican union is not a union—it is a dummy organization—Terrazas has served the growers" and consul's union will fail. On the other hand, the "Communist union" had "a better plan" for the workers, but with the wrong leadership.[97]

With melon picking set to start, the CAWIU sent carloads of Mexican organizers to the desert, but police monitoring the roadways arrested them. Recognizing federal officials had abandoned Glassford, agribusiness launched a campaign against the CAWIU and the ACLU. The civil liberties union piled on, calling for Glassford's removal due to his "open hostility to workers," and it requested federal marshals stop a "reign of terror" in the county.[98] When the conciliator began investigating vigilantism he received a death threat from a farm leader. Incensed, Glassford sent a scathing press release: "The valley is governed and controlled by a small group which, in advertising a war against communism, is sponsoring terrorism, intimidation and injustice." Glassford humiliated residents for permitting fascism: "it is time Imperial county awakened to the fact that it is part of the United States." Before departing, Glassford issued a press release: "a group of growers have exploited a Communist hysteria for the advancement of their own interests; that they have welcomed labor agitation, which they could brand as 'red,' as a means of sustaining supremacy by mob rule, thereby preserving what is so essential to their profits—cheap labor"; and county officials were "the principal tools of their machine." Farm leaders sharply criticized the report, and federal officials ignored it.[99]

The Cannery and Agricultural Workers Industrial Union's collapse in Imperial County in mid-1934 left Mexican, Filipino, white, and black fieldworkers without an independent field union. Anticipating a profitable upcoming lettuce harvest, in October, consul Joaquín Terrazas attempted to soothe working-class anger by convincing growers to increase wages by 30 cents an hour. Later that year, Consul Terrazas met with President Abelardo Rodríguez and the territorial governor in Mexicali. Based on President Rodríguez's history of suffocating working-class militancy in the Mexicali Valley, he was probably pleased with the consul's cooperation with agribusiness and with how the consul and the Mexicali arbitration board defeated the field strike.

Two weeks later, Consul Terrazas and Imperial Valley farm representatives established wages for the next cantaloupe harvest.[100] Soon thereafter, the government transferred Terrazas to another post, and his union quickly folded. Imperial Valley fieldworkers were left without union representation for a generation. In 1935, the arrival of thousands of white migrants, coupled with the influx of six thousand Filipinos, saturated the labor market and reduced wages. Suppression of Mexican unionism, scarce jobs, and rock-bottom pay convinced braceros to bypass the Imperial Valley or repatriate.[101]

Mexicanizing the Mexicali Valley

Following the failed 1934 strikes in the Imperial Valley, *agraristas* (agrarian militants) in the Mexicali Valley vented their fury by demanding jobs and petitioning the government for *ejidos* (communal farms). Unlike the small twenty-five federal colonies, five territorial colonies, and three private colonies already in existence, the ejidos were federal projects. While campaigning in September of that year, presidential candidate General Lázaro Cárdenas visited Mexicali to inspect the agricultural colonies. Upon his inauguration in November, Mexicans on both sides of the California borderland sent President Cárdenas numerous telegrams explaining their destitution and requesting assistance. Cárdenas dispatched officials to investigate conditions and settle grievances.[102] Cárdenas's intention to settle repatriates in Mexicali rested on the efforts of a previous president, Pascual Ortiz Rubio, who, within months of the 1930 Imperial Valley strike, sent Secretary of Communications and Public Works General Juan Andrew Almazán to the western borderlands to investigate the plight of repatriated and deported citizens. Almazán suggested purchasing "at any cost" the Colorado River Land Company's fiefdom and resettling there industrious repatriates. Almazán argued it was a matter of national security because Americans desired western Sonora and Baja California. The report criticized Asians and recommended their deportation. Almazán also urged colonization to increase usage of the Colorado River to strengthen Mexico's negotiating position before Boulder (later Hoover) Dam became operational.[103]

Hewing to the report's recommendations, President Ortiz Rubio issued a decree ordering the "Mexicanization" of Baja California, including purchasing land from foreigners and facilitating the colonization of expatriates. Responding to rumors of the US's desire to annex Baja California, Mexico's Congress strengthened its political grip on the peninsula by dividing it into two halves and granting them territorial status. Ortiz Rubio also commissioned a study to develop the isolated region. Countering Mexico's intent to tighten its grip in the region, in January 1931, an Arizona senator submitted a resolution to annex Baja California and the northwestern part of Sonora to acquire the Colorado

River. Although the annexation proposal went nowhere, Mexico's leaders were determined to populate the Colorado River delta.[104]

In 1935, Mexicali's farm economy emerged from its economic slump when cotton prices rose. Angling for jobs, Mexican repatriates and unions in Baja California petitioned the government to expel Asian laborers and tenants. Xenophobic screeds convinced the Chinese consul in Mexicali to repatriate the Chinese colony. Mexicali's Japanese colony soon departed as well. Throughout the year, bureaucrats traveled to Mexicali to study and promote colonization. In April 1936, Cárdenas finalized contract negotiations with the Colorado River Land Company to sell small farms to repatriates on long-term loans. The agreement, however, effectively locked out destitute braceros. In a national radio address five months later, President Cárdenas urged citizens to support his intention to build up the southern territory of Quintana Roo and the northern territory of Baja California.[105]

Disaffected *ejidatarios* (supporters of communal farms) responded to the government's disinterest in their plight by seizing farms and compelling the government to legitimize their actions. Their January 1937 land invasion, known as "El Asalto a las Tierras," is annually celebrated in Mexicali. Some four hundred ejidatarios invaded CLRC ranches and chased off the tenants. Troops hauled militant leaders to the stockade, but President Cárdenas immediately released them. Taken aback by the brazen landgrab, Cárdenas invited ejidatario representatives to meet with him in Mexico City. In the interim, militants occupied farms and told small landowners, tenants, and even long-time farmers living on the territorial communes to leave. When reports of these abuses reached Cárdenas, he ordered the territorial governor to send soldiers to quell the disorder.[106] Behind the scenes, influential political supporters urged the president to side with the land activists. After meeting with Mexicali ejidatarios, the president jettisoned the Colorado River Land Company's colonization agreement he signed the previous year and announced the government's intention to expropriate the CRLC's estate and redistribute it to the landless. The stunning news came in March from the chief of the agrarian department, who pledged to remain in Mexicali until the transfer of CRLC property to ejidos was complete. The government opened a land bank and an agrarian agency with an engineer to oversee the technical aspects of the operation. Emboldened by the expropriation announcement, ejidatarios threatened small Mexican farmers to leave. Frightened and confused farmers and tenants sent telegrams to the president begging for assurance their property rights were secure. In April, Cárdenas expropriated the bulk of the Colorado River Land Company's property (450,000 acres). When the CRLC protested, the president replied the company's 1936 contract violated the Agrarian Code.[107]

With the government unwilling to enforce martial law in the rural areas, Mexican farmers and tenants converged on the park across from Mexicali's

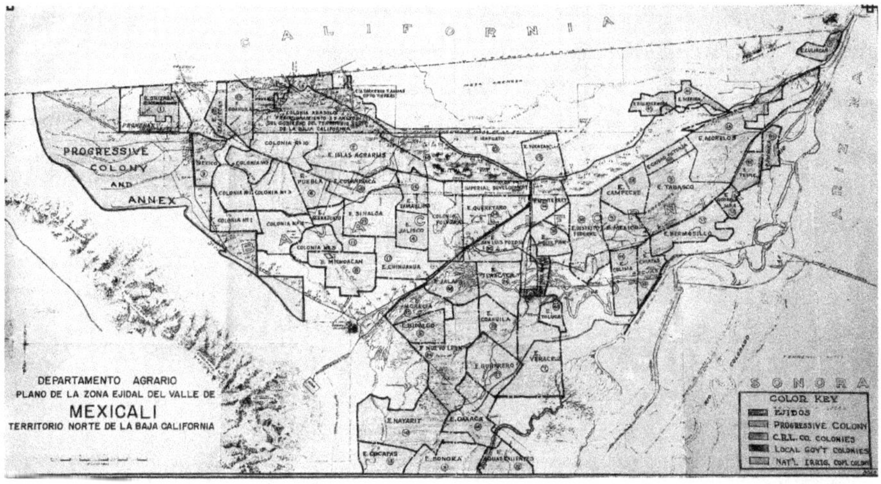

Forty-four ejidos planned for the Mexicali Valley, 1938. Courtesy Earl Linholm, "A Study of the Agrarian Revolution of Mexico as Applied to the Division of Land among the Peons in the Mexicali Region of Lower California" (master's thesis, University of Southern California, 1938).

territorial palace. For twenty days, about seven thousand protesters held a sit-down strike, refusing to leave until they met with the president. After tense negotiations Cárdenas modified his communal plan and exempted the existing colonies from the ejido program. Government officials also pledged compensation to farmers for damaged property.[108] Gambino Vásquez, the head of the Agrarian Department in Mexico City, personally oversaw the marking out of ejidos. By November, 4,730 male adults, along with family members, moved onto communal farms. By the end of 1937, the government established forty-four ejidos. In 1945, there were sixty-four ejidos with 5,353 families and thirty-six *colonias* covering 464,183 acres, an area roughly equal to the Imperial Valley.[109] In 1942, the American-Mexican Claims Commission awarded $3 million to the CRLC for its losses, despite the company's assertion that it had invested $12 million over four decades, and investors had yet to receive a penny for their investment.[110]

Shed Labor Disempowered

Across the border, Imperial Valley shippers turned their attention to subjugating packing shed unions. During the Great Depression, white shed workers, holding the benefit of whiteness and citizenship, struggled to maintain their status as the agricultural labor elite. Shed workers frequently went on strike to halt the slide in wages, with mixed results. In December 1934, seven months after labor conciliator Pelham Glassford departed, the Fruit and Vegetable Workers Union (FVWU), based in El Centro, asked shippers to negotiate wages and hours in the lettuce harvest. The union, which dominated packing in California and Arizona, wanted union recognition, a closed shop, and a wage increase. It insisted on a closed shop to prevent employers from hiring Japanese and Mexicans, a strategy used to lower wages and drive a racial wedge among the packers. When shippers stalled, the FVWU called a strike. The Brawley-based Wrap Pack Association walked off the job in support of the FVWU.[111]

Shippers hired high school and college students and hundreds of Mexican and white replacement workers. The strikers assaulted replacement workers and picketed the sheds. El Centro firemen used a water cannon to disperse the crowd. Police said women "give more trouble than the men. The feminine strikers are mostly young, wear overalls and sweaters, and are among the first to inquire at city and county jail when they hear any of their friends have been arrested." A FVWU leader exhorted Mexican field hands to join the walkout. One man said, "If you join us and stick with us, we've got the growers and shippers whipped." Employers warned field laborers to stay out of the dispute. While a federal conciliator met with both sides, replacement workers shot to death two picketers at a shed in El Centro. "The women hit the Wahl shed first," a scab testified, after a woman picked up a rock and said, "Let's get 'em." Four hundred strikers sprinted toward the shed flinging stones. Scabs responded with a hail of bullets, killing Eldrid Hamaker and Paul Knight, whose wife was in a hospital giving birth. On the day of the burial, a coroner's jury recommended a grand jury investigation into the killings. The district attorney packed the grand jury with "substantial" men who refused to charge the shooters for murder. One packer said, "The strike is lost. Our leaders are either in jail or have gone away."[112]

Shed workers flexed their muscle and lost. Shippers shoved a dagger in the FVWU by announcing plans to only hire local residents, which meant two-thirds of the union's members lost their jobs. A few months later, before the cantaloupe harvest, shippers offered a "non-strike bonus" for packers. The nonstrike bonus consisted of employers withholding a portion of the wages until the end of the season. Incensed, the Brawley-based Wrap Pack Association walked off the job. As the National Labor Board mediated, shippers placed a replacement worker in charge of the Wrap Pack union and recognized it as the shed bargaining agent. Agribusiness now controlled the county's field and shed workers. In July 1935, President Roosevelt signed the National Labor Relations Act, recognizing unions, guarantying collective bargaining, and mediating disagreements. Agricultural workers, however, were exempt from the law. Farmworkers were also exempt from the 1935 Social Security Act and the Fair Labor Standards Act of 1938. Shed workers lost every strike for the rest of the decade. The unionization of shed workers and teamsters was so intense, two white organizers were shot dead in 1938, yet no one was arrested for the murders.[113]

Conclusion

From Sacramento, a writer in 1935 penned an essay in the *Nation* asking: "Why is California the most fascist state in the Union?" and "Why are the agricultural workers of California today living under a virtual dictatorship?"[114] This chapter answers these questions by exploring the trajectories of power among a multiracial workforce; agribusiness; communists; liberals; local, state, and federal officials; and Mexican consuls and Mexican functionaries. Understanding the ideologies of power and control over the Colorado River, nature, and farm production explains the determination and ferocity of farm leaders to bring labor under total submission.

For four decades, field hands struggled to carve out a meaningful life in harsh environmental conditions, inadequate housing, and in the shadow of powerful agribusiness interests. During the colonization period, the Industrial Workers of the World attempted to organize workers in the Imperial Valley. Vigilantism and criminal syndicalism convictions wiped out the IWW. Furthermore, racial animus and job insecurity undermined interracial working-class cooperation. In 1922, braceros on both sides of the valley staged an enormous cotton and cantaloupe harvest disruption, resulting in increased wages. Meanwhile, in Mexicali, Chinese cotton pickers frequently went on strike until the 1920s, when braceros pressured employers to discharge foreigner workers and hire Mexicans. After the Mexican revolution, state-sponsored arbitration boards in Mexicali mediated workplace conflict and enforced employer-union agreements. During the profitable 1920s, Imperial Valley farmers paid high wages, but by the end of the decade, the proliferation of labor contractors and the rise of grower-shipper concerns and crop associations tipped the balance of power toward employers. In 1928, braceros united to end the contract labor system, raise wages, and obtain workplace reforms. Police repression prompted a state investigation and the regulation of labor contractors.

Plummeting wages and competition from American migrants during the Great Depression led to the valley's first interracial field walkout. During a 1930 lettuce harvest Mexicans and Filipinos demanded job reforms and higher wages. Hard-hitting police and antiunion sympathizers ended the conflict, but communist organizers rebuilt the field labor alliance and a few months later, during the cantaloupe strike, compliant judges imprisoned dissenters, and the Mexican consul encouraged repatriation to get rid of instigators. Police brutality and the arrest of the left-leaning union leaders lured the American Civil Liberties Union to investigate civil rights violations. The district attorney put the state's criminal syndicalism law to use for the first time to convict communist union organizers. The 1930 field revolts set the battle lines for a decade of labor unrest. Field hands faced incredible odds when matched against the autocratic rule of agribusiness. In addition, workers remained divided by race and class. Agriculture's aristocracy—shed workers—consistently ignored or manipulated fieldworkers. The mayor of Brawley reflected the attitude of the farm elite toward fieldworkers: "White men are no good in our business. We like our Mexicans; they don't complain, and they don't ever do any organizing."[115]

The year 1934 was a turning point in employer-worker relations. Mexicans, whites, blacks, and Filipinos joined the Cannery and Agricultural Workers Industrial Union. The CAWIU, in an alliance with lawyers from the American Civil Liberties Union, grappled with a farmer-sponsored union created by Mexican consul Joaquín Terrazas. The consul's undermining of the CAWIU revealed the tension between braceros' belief in their labor rights and the Mexican government's objective to maintain social harmony, the employment of its citizens abroad, and good relations with its northern neighbor. Unlike in the Mexicali Valley, where unions wielded enormous power and compelled employers to negotiate contracts, in the Imperial Valley, transnational subjects had no such rights. Within this context, the CAWIU engaged police and vigilantes, stern judges, and functionaries of the United States and Mexico to determine if fieldworkers had a right to shape their economic destiny. Agribusiness formed the Associated Farmers and the Anti-Communist Association to smother all challenges to their authority. The CAWIU's fate was sealed when shed workers and a federal conciliator refused to support the union. Agribusiness disempowered shed workers the following year in a deadly lettuce strike. Consequently, strike-killing measures emasculated agricultural unions in the Imperial Valley. Yet the agency of transnational migrants bore fruit across the border when ejidatarios secured the Mexicanization of the Mexicali Valley in 1937. Decades of unrelenting activism culminated in President Lázaro Cárdenas's expropriation of the Colorado River Land Company's vast landholdings and the establishment of forty-four ejidos.

Conclusion

> A struggle to the death from year to year
>
> David Bromley

In November 1930, Imperial County residents celebrated at the county fair. They had cause for joy since the valley was one of the nation's few "white spots" of economic vigor. Just a year earlier, President Herbert Hoover signed legislation to build the All-American Canal. To honor the moment, David Bromley wrote a poem titled "The Last Frontier." Falling within the literary trope of desert settlement, its central theme centered on the struggle with, and conquest of, a desolate and deadly desert:

> Grim land of staring death and desolation,
> The march of progress halts before your door,
> Man drives you back a mile in desperation,
> But on the morrow yields to you once more
> A fight with drought by means of irrigation
> A struggle to the death from year to year,
> The vanguard of a great and mighty nation,
> Fights nature here on this, the last frontier.[1]

Embodying the Darwinian thesis of natural selection, the poem merged the triumph of American civilization with the transformation of an arid landscape into verdant farms. "The weak you cast aside crushed down and weeping" and "the strong you gather quickly to your breast" to make a desert Eden. Nature challenged "man" but gave way to the chosen ones. Man despised nature, but once he conquered it he respected it.[2] The nationalistic image of the select, the masculine, ordained by the Almighty to subjugate the nation's last frontier, validated white settlers' struggles and sacrifices. The belief in colonizers as the anointed ones had roots in the psychology of power and control guiding the Imperial-Mexicali Valley's history; thus each chapter has explored the mechanics of acquiring, consolidating, wielding, and resisting power.

This study reveals a systemic drive to dominate and manipulate the environment, perfect desert irrigated farming, construct a racialized hierarchical labor force, and police a racially segregated society. Molding water, landscape, labor, and social relationships reflected ideologies of power and the desire to dominate. Imperial Valley settlers and the Colorado River Land Company in Mexicali believed the desert valley was a calamitous world, with the ever-present threat of a "rampaging" river, irrigation droughts, insects, "pests," and the vicissitudes of the market. Agribusiness dealt with these uncertainties by seeking to overcome all challenges, including eliminating "pests," silt, and alkali. An illustrative example of vanquishing the opposition is seen in the campaign to wrestle the Colorado River from Mexico and to build an All-American Canal. Settlers and outside investors transferred the psychological desire for controlling nature for agricultural profits into defeating political and economic dissidents and obtaining an abundance of migrant laborers and then dividing them by race and class. Transforming the desert into an agricultural breadbasket entailed cooperation among corporate interests, banking, and local, state, and federal government functionaries. We see these cooperative relations, first expressed by Harry Chandler and H. H. Clark of the Colorado River Land Company, in procuring braceros after World War I. Agricultural interests created crop and labor associations as a natural progression of capitalism while at the same time denying field hands the right to organize unions for collective bargaining.

Fieldworkers in the United States had few advocates and lacked representation in the halls of power. Nevertheless, while federal officials in Mexico City and Washington, DC, allowed the procuring of international migrant labor, reformers in California demanded the enforcement of school attendance, wage, and housing laws. Although growers mostly agreed to ban child labor, they declined to enforce sanitation and housing laws. Braceros in Mexico lacked sanitary living and working conditions, but they had access to power unaccorded workers in the United States. Mexico's 1917 Constitution granted workers the right to organize unions, negotiate with employers, strike, petition the government, and demand state-approved arbitration. Thus while Imperial Valley farmers crushed field unionization, braceros across the international line formed unions, staged walkouts, negotiated with employers, and compelled the government to enforce a law guaranteeing 80 percent of jobs for Mexicans.

For four decades field and shed workers empowered themselves with a repertoire of options, ranging from daily acts of resistance, crew walkouts, union organizing, and in extreme circumstances, by striking to increase wages or improve working and living conditions. In the Mexicali Valley, landless braceros squatted on foreign-owned farms and protested against the employment of foreigners. Chinese and bracero cotton pickers frequently refused to harvest crops until employers increased wages. Each defiant act revealed how workers reshaped the trajectory of their working and living conditions as they pushed to reorder power relations on the farm and in the packing sheds. We see the evolution of increasingly powerful coalitions clawing for control over decision making during the great field strikes of 1934, when an interracial alliance of migrants, liberal lawyers, and communist organizers waged a titanic battle against agribusiness, police, judges, vigilantes, and functionaries of the United States and Mexico. Drunk with power, agents of industrial agriculture unleashed a reign of terror that ended in total victory for employers. White shed workers met the same fate the following year.

In the mid-1930s, California agribusiness created the Associated Farmers, based on the model of vertical integration established in Imperial County. Carey McWilliams described the Associated Farmers: "Carefully coordinated from the top, the organization network functions with clocklike precision, policing entire crop industries, enforcing uniform decisions, holding recalcitrant employers in line."[3] A spokesperson described the fascist mentality after a violently repressed harvest walkout: "These men are accustomed to meet their problems firmly, and whether or not others agree with the methods they chose in meeting this particular one, nevertheless it must be admitted that they did meet it and that they met it with determination."[4] As police and the courts punished and banished political activists and labor organizers, white middle-class moral guardians watched for social subversives who challenged racial segregation and moral precepts.

As a borderlands case study, incorporating the history of the Mexicali Valley illuminates efforts to share and fight over the Colorado River, organize crop associations, procure bracero migrants, and patrol the towns and the international border to root out economic, political and social dissenters. As agribusiness in the Imperial Valley disempowered migrant fieldworkers and packing shed laborers during the Great Depression, braceros crossed the border and manifested their own authority by seizing foreign-owned farms in the Mexicali Valley. Agrarian militants persuaded President Lázaro Cárdenas to expropriate the Colorado River Land Company's property and reconstitute its vast estate into ejidos. In sum, this project has demonstrated how the contest for power and control drove the historical narrative of this agricultural dynamo straddling the California–Baja California borderland.

Afterword

> We can't get work, we can't get money and they don't give grants.
>
> Sarah Meza, El Centro, 2009

What has a century of intensive irrigated industrial agriculture wrought? What is the legacy of environmental degradation? What has the destruction of field unions and the unbridled support of Mexican migrant day labor bequeathed? This afterword answers these questions while describing the stunning reversal of the Imperial Valley's reputation and political power over the previous three decades. Once lauded for its agricultural output and its lion's share of the Colorado River, today the Imperial Valley is infamous for its flagrant water waste, unsustainable farming practices, feeble economic output, endemic poverty, and debilitating pollution. Despite its diminished power and tarnished reputation, this desert valley remains at the center of economic and environmental controversy as the Southwest faces profound climate change and water scarcity challenges. Following an overview of contemporary Imperial Valley, the next section updates the themes in the previous chapters, tracing the major events and trends from World War II to the present. A third section discusses the region's intractable socioeconomic-ecological challenges. The book concludes with six recommendations for refashioning the desert's socioeconomic-ecological structure.

Imperial Valley

The following statistics are indicative of the twentieth century's legacy of conquest and manipulation of nature; intolerance of social, political, and economic criticism; and the ironclad control of farm labor relations in Imperial County. In 2010, Latinos made up 81 percent of the county's 174,500 residents, with 32 percent of the population born abroad. The valley has the inglorious distinction of ranking highest or lowest in negative state and national socioeconomic indicators. It is the state's poorest county, with jobless rates fluctuating from 19 to 32 percent, the highest nationally. After unemployment reached 32 percent in October 2009, the *Associated Press* ranked Imperial County as the most economically stressed county in the United States. From 2009 to 2011, El

Centro, the county seat, had the highest joblessness of any town in America. With the economy in a tailspin, the county placed first in the state for bankruptcies. Unsurprisingly, the county ranked last in median household income in the first decade of the twenty-first century. In 2010, for instance, 36,000 families (23 percent of the populace) lived in poverty, and one-third of the population received public assistance. Poverty is linked to other disreputable socioeconomic indicators. For instance, the valley has the lowest literacy in the most illiterate state in America.[1]

Unrelenting despoliation of nature for commercial agriculture and livestock raising in the twentieth century resulted in millions of pounds of chemical poisons sprayed from airplanes and mixed into irrigation water, fouling water, land, and air. Emblematic of this catastrophe are the drains of the New and Alamo Rivers, which flow south from Mexico to the Salton Sea. The New River, which crosses from Mexicali into Calexico, contains pathogens and contaminants from Mexicali's effluent and mixes with salty water from field runoff, selenium, and industrial agricultural chemicals. The New River, the most polluted waterway in the United States, oozes past Brawley. In 2013, a report identified Brawley as the most polluted town in California, a ranking determined by a plethora of environmental hazards.[2] Evidence of the crippled desert ecosystem is found in health data. Imperial County's air quality, the worst in the state, impacts one-fifth of its residents, who suffer from child asthma and other respiratory ailments. In 2009, the American Lung Association ranked El Centro as the tenth worst city in the United States for ozone healthfulness. Unsurprisingly, the valley leads the state in adolescent obesity and diabetes. State and federal air quality agencies found Imperial County in violation of air quality standards and forced the county to implement stringent regulations, which hinders economic activity.[3]

Mexicali Valley

Across the border in working-class Mexicali, the capital of Baja California and home to over a million residents, leaders charted a different course from the Imperial Valley. Mexicali's economic growth spiked with high cotton and commodity prices during the Korean War. Ejidatarios expanded acreage by a factor of four between 1937 and 1960. Increased farming acreage lured workers to the region, increasing Mexicali's growth from 45,500 in 1940 to 192,500 in 1957. Mexicali's farming boom ended in the 1960s when water limitations mandated in the US-Mexico water treaty of 1944 took effect. The treaty limited Mexico's share of the Colorado River to 1.5 million acre-feet, less than half the amount granted to the Imperial Valley.[4] With less acreage under the plow, in the 1980s Mexicali diversified the economy with 244,000 maquiladora assembly jobs, border commerce, and agricultural production, fueling urban growth and economic and social stability. Unlike other Mexican border cities, Mexicali avoided vicious drug gang wars and indiscriminate murders of women. Residents cross the border to shop in Calexico and El Centro, buoying those towns' anemic economies. Living in the shadow of the United States, however, is a double-edged sword. US immigration policies have created a humanitarian crisis of homeless deportees. Every day, 100 to 140 expelled Mexicans walk through an eighteen-foot-tall metal fence separating Calexico from Mexicali. In 2011 alone, 63,000 deportees returned through this port, often in the middle of the night. Hungry, penniless, and sometimes shoeless, the downtrodden make their way to El Hotel del Migrante Deportado, a rundown structure where America's discarded workers sleep on concrete floors and subsist on beans.[5] This synopsis of modern-day Imperial-Mexicali Valley can best be understood with a brief summary of major events and trends since World War II.

Ben Yellen's Righteous Cause

Industrial farming recovered from the Great Depression with demand for food and fiber during World War II. Food and fiber production during the Korean and Vietnam wars sustained the valley's economic prosperity. The Interior Department's water and agricultural programs convinced federal officials to disregard the Imperial Irrigation District's (IID) noncompliance of the reclamation act's 160-acre limitation and its farm residency requirement after completion of the All-American Canal in 1942. The government's cooperation with corporate agriculture paralleled its silence on the IID's dubious water rights claim to the Colorado River. The cozy relationship between the IID and the federal government took an unexpected turn when the Department of the Interior belatedly sued the IID in 1967 to enforce the 160-acre limitation to partake in federal irrigation projects. Ben Yellen, a Jewish physician who relocated from Brooklyn, New York, to Brawley in the mid-1940s prompted the government's startling legal challenge. The irrigation district undervalued farm property to reduce taxes, and it increased electricity rates to stabilize the price of cheap farm water. Incensed by the IID's farm subsidies, Yellen launched a one-man public awareness campaign against agribusiness injustices. When a judge ruled against the government in the 160-acre limitation case, Yellen and 123 jobless and landless Mexican Americans sued the Department of the Interior for nonenforcement of the reclamation act's residency requirement. The petitioners sought to purchase farms from absentee landowners.[6]

While Yellen's case wound through the federal courts, the irrigation district negotiated a settlement adjudicating its water rights. The US Supreme Court issued its decision in *Arizona v. California* (1963), which settled a long-running legal battle between the two states over the Colorado River. In doing so, the court named the secretary of the interior as the river "water master" to regulate the waterway and to distribute water during shortages. The court also ordered California, Arizona, Nevada, and Native American tribes to settle claims of "present perfected rights" to the river under state or federal law. Farm owners in the Imperial Valley knew they had a flimsy claim to the river, for as this study has conclusively demonstrated, irrigation water from the Colorado River never entered the Imperial Valley with a legal filing under California law. In fact, the Imperial Valley's water claim lacked validation from a California court or a water right from Congress. Rights to the diverted portion of the river were held by the Mexican government through its 1904 concession/law and reaffirmed in 1910 with the California Development Company. A claim to those water rights passed to the US government when it built Imperial Dam and the All-American Canal. The IID's powerful allies at the state and federal level distorted the historical record and pushed for court validation, which eventually came in a 1979 Supreme Court supplemental decree. The court granted the IID 2.6 million acre-feet per year plus more in surplus years. The decree upheld the fictional claim that the Imperial Valley held water rights before the Boulder Canyon Project Act (1928), the law approving construction of Hoover Dam and the All-American Canal.[7]

A year after granting the Imperial Valley water rights, the Supreme Court ruled against Ben Yellen and the 123 ethnic Mexican petitioners. The Supreme Court used the same flawed logic as its 1979 supplemental decree in rendering its decision: that the IID had a present perfect right to the Colorado River under state law before enactment of the Boulder Canyon Project Act, making landowners exempt from the reclamation act's residency and 160-acre provisions. The decision disregarded a host of reclamation acts to the contrary and barely acknowledged that water entered the valley from Mexico. The decision downplayed multiple contracts approved by voters between the IID and the Interior Department that specifically referenced government projects were subject to the reclamation act. Two years later, Congress raised the minimum acreage under federal irrigation projects from 160 to 960 acres and eliminated the residency requirement. Finally, in 1987, the Bureau of Reclamation gave corporate agriculture all the water it wanted regardless of the landholding size.[8] Due to these changes, today there are some 450 farmers in Imperial County, and the average farm is 957 acres. By way of comparison, in 1929 there were 4,769 farms (landowners operated 1,860 and tenants handled 2,909).[9]

Ben Yellen's righteous cause for socioeconomic justice for farmworkers failed because the Supreme Court ignored the original intent of subsidized federal irrigation projects. The court also upheld the reclamation movement's original intent to promote white supremacy in the farming districts of the American West. In addition to validating law breaking, these legal and legislative decisions perpetuated racial and class inequities by circumscribing Latino opportunities. The government's indifference in enforcing the reclamation laws, coupled with the 1980 Supreme Court decision, and subsequent congressional legislation affirming these decisions, engendered economic stagnation in the Imperial Valley. With tenants running 60 percent of the ranches, a huge portion of agricultural profits leaves the valley to enrich outside investors, reducing the amount of money circulating locally.[10] Pervasive absentee ownership explains the miniscule commercial sector; broad impoverishment; and the annual exodus of its industrious, intelligent, and talented youth.

Agribusiness and Racial Politics

Paralleling its racial water politics, the federal government cooperated with desert farmers to uphold white supremacy by ending Japanese domination of the profitable fruit and vegetable industries. In the weeks after Japan attacked Pearl Harbor in December 1941, there were several physical assaults on Japanese in the valley and an unsolved murder of a Japanese married couple on their ranch. In February 1942, the county Board of Supervisors passed an ordinance aimed directly at the Japanese. It ordered enemy aliens to register and forbade enemy aliens from working in agriculture in any capacity except as laborers. At the time, there were 1,583 ethnic Japanese who owned or leased 212 farms. Since most were immigrants barred from citizenship, the ordinance affected almost all the Japanese farmers and shippers. Violence and a direct threat on their livelihood compelled Japanese to sell their property before they entered a war internment camp. In many instances, white landlords sold crops farmed by Japanese and used every imaginable legal trick to acquire Japanese farms and property. A wartime government report noted, "The Valley has never had a tradition of law and order" to explain the serial violation of civil rights. Near the end of the war, farm leaders held mass meetings to rally against the return of Japanese evacuees. The government concluded, in light of racist hysteria and inadequate protection, it did not recommend relocating Japanese to the valley. Only a small number of Japanese returned and continue farming today.[11]

Today, farmers grow over forty different crops, raise livestock, and farm aquatic products. Vegetables and melons encompass almost half the gross profits, while livestock produces about one-third, and field crops generate 20 percent of economic output. The vegetable and livestock industry is not as profitable as in past decades. Even with 450,000 acres, ideal weather, bountiful and subsidized water, government crop payments, and inexpensive and plentiful labor, Imperial County's agriculture and livestock industry never exceeded $2 billion in gross output annually during the last decade. The local economy continued its downward spiral when the beef processing plant in Brawley closed in April 2014, resulting in the loss of 1,300 jobs. The plant's closure will result in an 80 percent decline of the $500 million a year cattle industry. Economists estimate a reduction of another 1,040 jobs that supported the beef industry. The devastating news meant the loss of $102 million to the local economy and will raise unemployment another 5–6 percent.[12] As discussed in chapter 3, Imperial County farm property owners and tenants benefited from an array of state and federal technical and financial support programs. In the twentieth century, agricultural and livestock producers reaped billions of dollars in subsidies, technical support, market advice, and tariff protection. According to the Environmental Working Group's Farm Subsidy Database, from 1995 to 2012, Imperial County resident landowners received over $112 million in cash crop subsidies, with an undisclosed amount received by nonresident owners.[13]

The subjugation of nature, a theme of previous chapters, continues unabated. Chemicals to maximize production are sprayed from planes and injected in irrigation water. The ideological consistency of the conquest of nature is reflected in the County Agricultural Commissioner's reports. For instance, in 1949, farmers took out permits for three million pounds of insecticides. The following year, the amount of insecticides doubled.[14] In the first decade of this century, farmers received 15,000 permits annually to eradicate a wide assortment of "pests."[15] Outside the valley, the Department of Agriculture's Wildlife Services is the primary "killing agency" according to the *Sacramento Bee*. The newspaper describes indiscriminate, illegal, and inhumane poisoning, trapping, and aerial gunning operations that eliminated millions of birds, almost a million coyotes, numerous mammals, over 1,100 dogs, and ten humans. Similar to the Imperial County's extermination efforts, the government's lethal wildlife control program receives scant publicity.[16]

Braceros

As racists expelled Japanese from farming in 1942, agribusiness sought federal assistance to push white migrants out of the fields. During World War II, farming interests joined other industries to lobby the government for seasonal Mexican laborers. As a result, the United States and Mexico created the Bracero program to permit unattached men to work temporarily in the United States. Agricultural leaders persuaded the government to continue the program until 1964. Farmers used the program as cover to inundate the county with undocumented Mexicans. Employers lowered wages with a combination of braceros, unsanctioned border crossers, and Mexicali day laborers. Field wages remained constant at 70 cents an hour from 1952 to 1959, while nonagricultural pay rose 52 percent. Agribusiness's illegal activity resulted in a 1945 government report estimating 10,000 undocumented field hands in the valley, whose presence angered American job seekers. The report, unsurprisingly, did not spur the government to enforce immigration or labor laws.[17]

In the late 1950s, over 14,000 braceros made up 80 percent of the county's field labor force. Blessed with an overabundance of workers, farmers routinely violated the Bracero program's mandate for health care, the prevailing wage, and adequate food and housing. This continued the tradition of labor abuses documented in chapter 4. When jobless US citizens picketed for jobs in 1960–61, the police, deputized growers, and federal officials defeated workers in much the same way this alliance crushed strikes during the Great Depression. A total of 4.5 million braceros cycled through the program before critics canceled it. In 1979, César Chávez's United Farm Workers (UFW) attempted to unionize Imperial Valley's fieldworkers. The UFW strike crumbled after a farm employee who shot to death a picketer went unpunished. Growers have thwarted field union organizing ever since.[18]

A century of relying on foreign migrant workers has yielded a harvest of poverty and abuse. Currently, the valley's farm and livestock industries employ 24,000 to 31,500 people. Most of the county's 6,600 day laborers reside in Mexicali, where they line up at the border crossing at 1:30 a.m. to stand in line for two to three hours to enter the United States. Once over the line, migrants wait to be selected by a farm labor contractor, who transports them by bus to the fields. Laborers are not paid for traveling up to an hour to far-flung fields. After working from 6:00 a.m. to 2:30 pm they take home about $59. The labor contract system is rife with abuses. Workers frequently pay the contractor a kickback for hiring them and receive no health or retirement benefits. In addition, laborers endure sexual harassment and are exposed to farm poisons. Corporate agriculture accepts no responsibility for this unethical system, while the government refuses to intercede. Ignoring an army of local job seekers, farmers vigorously lobby for H-2A visas for foreign workers.[19]

Water Transfers and Environmental Justice

Despite its miniscule population and meager economic output, the Imperial County remains at the center of the Southwest's economic future due to its enormous Colorado River allocation. Yet, ironically, in 1980, the year the Supreme Court defeated Ben Yellen's bid for farmworker justice, the Imperial Irrigation District's domination of river politics began a three-decade slide when its former allies and enemies joined forces to downsize the water behemoth. The outcome ushered in a new era of large-scale farm-to-cities water transfers in California. The dispute began in 1980 after a farmer filed a complaint with the state claiming the Imperial Irrigation District's wasteful water practices flooded his property. After a lengthy investigation and hearings examining the valley's antiquated irrigation system, the state ordered the IID to find an urban partner willing to modernize its water distribution system in return for transferring the conserved water. In 1989, the IID and the Los Angeles–based water distributor finalized a deal.[20]

The water swap unleashed the water transfer genie from the bottle to terrorize the Imperial Valley. Faced with increasing urban population growth in the Southwest, the Interior Department reversed a century of federal benevolence toward California. The Colorado River "water master" set in motion a plan to wean California, principally the Imperial Valley, off its overuse of the river to allow other states an adequate water supply. State and federal water agencies no longer tolerated the valley's leaky canals and flood irrigation for marginally profitable crops. In the 1990s, the federal government instructed the irrigation district to partner with an urban entity willing to pay for irrigation conservation in return for the water saved. Bewildered that former allies were now adversaries, the IID reluctantly negotiated a deal with San Diego for the largest farm-to-city water transfer proposal in US history.[21] After years of heated opposition, lawsuits, and threats from the Interior Department to reduce the Imperial Valley's entitlement, the IID approved a landmark Quantification Settlement Agreement (QSA) in October 2003. Instead of bringing "peace on the river" for seventy-five years, however, the QSA sparked a decade of legal challenges to invalidate it.[22]

Controversy over the QSA centered on the water transfer's impact on the Salton Sea, the 376-square-mile agricultural sump on the northern end of Imperial County replenished from irrigation runoff and effluent in the New and Alamo Rivers. With the elimination of 95 percent of California's wetlands in the last century, the Salton Sea provides habitat for several endangered species and is a vital stopover for migrating birds along the Pacific Coast. Over four hundred resident and migrating bird species—the largest documented number in North America—nest there. Before the QSA, the saltine lake periodically experienced massive fish and bird die-offs from disease and organic decay from hydrogen sulfide irruption, causing a distasteful odor to waffle throughout the region. The water transfer nudged the putrid Salton Sea another step closer to a biological collapse. With salinity levels rising 20 percent from June 2003 to 2011, the lake's salinity level is 33 percent higher than the Pacific Ocean. For half a century, regional officials and natural resource experts recognized how excessive salt levels imperiled the lake. An estimated three to four million tons of salt enter yearly, and experts estimate the sea currently holds 450 million tons of salt. When the salinity level reaches the breaking point from reduced farm-water runoff, experts predict colossal fish die-off and the loss of critical food and habitat for wildlife.[23]

In addition to the negative impact on wildlife, a smaller Salton Sea will exacerbate the region's unhealthy air quality. Reduced inflows will expose 134 square miles of shoreline consisting of finely textured silt, mixed with salt and laced with selenium and pesticide and insecticide residue from a century of industrial farming. The exposed seabed will release its toxic brew into the airshed in enormous dust storms, like "a wall of dirt 20 stories high rolling across a populated area," as Imperial County Air Pollution Control District officer Brad Poiriez remarked, to foul fields and aggravate air quality. Without a restoration project, Tim Krantz, an environmental studies professor at the University of Redlands, predicts Imperial County "would have the worst air quality situation in the Western hemisphere."[24]

The QSA included a pledge by Southern California water users and the state to study and pay for Salton Sea biological mitigation. In 2009, the California department of water resources selected a Salton Sea restoration project costing $8.9 billion, a mindboggling sum the state legislature refused to consider. Imperial Irrigation District general manager Kevin Kelley excoriated legislators for their unwillingness to uphold the state's contractual obligation to fund a sea restoration project: "No one believes it will ever be funded." A decade after signing the QSA, "the Salton Sea is no closer to a viable solution." Alarmed at the state's and federal government's refusal to commit funding for a restoration plan, the IID and the county, at odds over the water transfer, reached agreement to cooperatively gather funding for the ailing lake.[25] A long list of opponents, including environmentalists and the Imperial County Board of Supervisors, sued the irrigation district to modify or invalidate the QSA. Critics claimed the agreement did not adequately mitigate the profound public health danger. County leaders also objected to projected reductions in economic productivity. Large farm owners sued, claiming they owned the valley's water rights, not the IID. The landowners lost every lawsuit, repudiating the flawed 1980 Supreme Court ruling implying the landowners, and not the public, were the beneficiaries of the irrigation district's water right. A court of appeals validated the QSA in July 2013.[26]

Notwithstanding the choreographed complaints from the farm community, the water transfers brought a fortune to the desert. In return for conserved water, urban water districts pledged to pay the IID $7.87 billion from 2007 to 2047. In a century-old tradition of pervasive farm subsidies, the water transfer perpetuated a legacy of cheap farm water. The IID took the financial windfall and set aside funds to keep water prices among the lowest in the American West. For the first fifteen years of the Quantification Settlement Agreement, the water district must conserve water by paying landowners millions of dollars a year not to farm 30,000 acres. In the world of paper farming, landowners paid the irrigation district $20 for every acre-foot and buyers sold the same water back to the IID for $125. Beginning in year sixteen of the seventy-five-year deal, the QSA obligated the irrigation district to pay farm landowners and tenants for installing water conservation measures such as sprinkler systems and concrete-lined canals. Yet eleven years after signing the

Quantification Settlement Agreement, the irrigation district has signed only a handful of water conservation contracts. Meanwhile, transfer income has paid for a platoon of lawyers and consultants, and to handsomely compensate an expanding bureaucracy of administrators and staff drawn from the agricultural sector's family members and acquaintances.[27]

The QSA reinforced local inequities by benefiting farmers and agricultural-related businesses, while mostly ignoring the larger society who supported agriculture in myriad ways. The IID established a program to compensate farm service businesses for the reduction in farming. In contrast, the IID handed out a fraction of the water transfer revenue to the broader community. The water district distributed on a competitive basis small grants for food, worker training, and job creation or retention programs.[28]

The Salinity Crisis and the Cucapá

Although the US-Mexico water treaty of 1944 denied Mexico enough water to irrigate the upper Mexicali Valley farms and the Cucapá homeland in the estuary farther south, the treaty's water limitation on Mexico did not initially take effect. Mexicali's 1950s agricultural boom came to a screeching halt in 1961 after the Bureau of Reclamation dug a canal in Yuma, Arizona, to allow salt from field runoff to drain into the river. The hypersaline water, unfit for human consumption, flowed downstream and ruined crops in the Mexicali Valley. Within weeks, 35,000 Mexicans protested in front of the US consulate in Mexicali. In 1974, the United States attempted to mitigate the damage with a billion-dollar desalination plant and a drainage canal away from the Mexicali Valley.[29] The salinity imbroglio impacted Mexicali farmers, yet it did not involve the indigenous Cucapá living in the lower Mexicali Valley delta. American efforts to capture more of the river, however, soon posed the greatest threat to Cucapá survival since the 1905–07 floods. When Arizona's Glen Canyon Dam became operational in 1963, Americans impounded the floodwaters of the river. Farmers in the Mexicali Valley took what they needed, reducing the once mighty river to a trickle before it disappeared altogether. Without a river flowing through the delta to the gulf, the immense wetlands in the lower Mexicali Valley dried up, turning vast stretches of the estuary to desert and salt flats.[30] Germán Muñoz, a Cucapá who witnessed this environmental disaster, saw the last dolphin sixty miles upriver from the gulf the year of Glen Canyon Dam's completion. Instead of the Colorado River meandering past the Cucapá village on the way to the gulf, the Cucapá live alongside a polluted stream of sewage from Mexicali and pesticides and salt leached from farms.[31]

The Cucapá's sorrowful plight stems from racist federal policies conceived in Washington, DC, and Mexico City during the Great Depression. President Lázaro Cárdenas's expropriation of foreign-owned lands in the Mexicali Valley negatively impacted the Cucapá, who lost jobs to Mexican settlers. Some moved to Arizona, while others trekked south of the ejido zone of development where they eked out a difficult living as day laborers. Almost a century earlier, the US-Mexico border cleaved the Cucapá homeland in half, dividing the tribe on both sides of the borderlands. During the 1930s, this imaginary line became a hard border when US immigration policies prevented the Indians from traveling back and forth. Many of the Cucapá who remained in the Mexicali Valley married Mexicans and formed bicultural families. Separated by an international border and confronted with US immigration policies that made it difficult or dangerous to cross the line, the tribe descended into bitter factionalism between the American Cucapá and their Mexican cousins. Although Mexico granted the Cucapá communal rights to their homeland in 1976, they lacked water rights to irrigate it. A flood in 1993 wiped out their tourist recreational campground, forcing them to survive on chickens, gardens, and fish from the contaminated stream. For the past two decades, American and Mexican activists and natural resource experts campaigned to rejuvenate the delta and help the Cucapá.[32]

While southwestern states reap the benefits from the Colorado River, Mexicans suffer from Americans' overuse of the river. Today, three million residents in northern Mexico rely on the river. Years of overtaxing the river harmed the delta's ecology and resulted in Minute 319, a landmark revision of the 1944 US-Mexico river treaty. The 2012 agreement allows Mexico to store some of its water allotment in Lake Mead to maintain the reservoir's elevation during a drought. Mexico also agreed to be lashed to the same river-use criteria as southwestern states, using surplus water when available and taking less during shortages. In return, US cities pledged to repair severely damaged irrigation infrastructure in the Mexicali Valley from a 2010 earthquake. The United States kept the water saved from these projects. In a nod to US and Mexican environmentalists, Minute 319 stipulated for the United States to provide water for rejuvenating the desiccated delta. In sum, urban southwesterners paid farmers on both sides of the Imperial-Mexicali Valley to modernize their irrigation system in return for the conserved water. Interestingly, while California refuses to fund Salton Sea restoration, Minute 319 sends water to reinvigorate the Colorado River estuary.[33]

A month after signing Minute 319, the Bureau of Reclamation released a comprehensive report on the future of the Colorado River and the water allotment system that serves seven states, forty million US residents, three million people in Mexico, twenty-two Native American tribes, 5.5 million acres of farm and ranch land that grows 15 percent of the nation's food, fisheries, eleven National Parks, seven National Wildlife Refuges, four National Recreational Areas, a $26 billion recreation industry, and hydroelectric power for cities. Modeling projections estimate that by 2060, a combination of drought, climate change, and population growth of between nine and thirty-six million people will outstrip the amount of water in the river. Drought is the most immediate threat, with the government predicting river cutbacks in 2014. The alarming report, and a river advocacy group's designation of the Colorado River as the most endangered waterway in the nation four months later, sent shock waves through the Southwest. In response, the Bureau of Reclamation hosted a major conference in San Diego for all river users to discuss how to address the impending calamity. Although representatives from the Imperial Irrigation District attended the conference, they vigorously protected the valley's water from its many rivals. Tina Shields, the irrigation district's river resource manager, quipped: "We like to farm. I don't think anybody down here is going to volunteer" for more water transfers.[34]

In hindsight, the Supreme Court's 1979 supplemental decree granting the Imperial Irrigation District present perfected rights to 2.6 million acre-feet and its 1980 ruling against Ben Yellen and landless petitioners solidified the irrigation district's water rights and legalized large-scale absentee farm ownership. To get around these obstacles and to compel the IID to transfer water to cities, state and federal governments challenged the IID's inefficient water distribution and farmers' wasteful irrigation practices. To date, this strategy yielded gargantuan water swaps, but at great cost to the Salton Sea and to the air quality and economic well-being of Imperial County.

Final Thoughts and Recommendations

How is it that a region conducive to year-round farming, with a half-a-million acres of rich soil, plentiful and subsidized water, possessing abundant, inexpensive, foreign nonunion labor, supported by compliant state and federal officials, fortified by unchallenged political, judicial, and police power, has bred widespread poverty and unhealthiness? This book argues that irrigated agribusiness imperialism created this state of affairs. The Imperial Valley's intractable problems are rooted in a cluster of beliefs in the conquest and subjugation of nature and labor to maximize farm profits. Agribusiness's supremacy over water, land, and labor relations has decimated the desert's ecology, violated the nation's land and water resource laws, trampled on the civil rights of critics, exploited workers, and committed egregious crimes against labor organizers and picketers, including murder. The taxpayers, the benefactors of irrigated desert farming, should rightly ask why they should continue tolerating and subsidizing agribusiness's economic mismanagement and deplorable environmental stewardship.

The region has reached a critical stage in its ecological life, and the socioeconomic-ecological structure must be reconstituted by enacting the following six reforms: first, fallout from a reduced Salton Sea will exceed federal air pollution standards and violate laws protecting wildlife and endangered species. Landownership beneath the saline lake is shared by the IID, California, and the United States, and all three entities cooperated for a century to use as much water as possible to deprive Mexico and other basin states of a fair share of the river. California and the nation's treasury have reaped a bonanza from the Imperial Valley's claim to the Colorado River. Furthermore, for half a century local, state, and federal officials knew the Salton Sea's ecology would collapse with hypersalinity, yet no one claimed culpability. Therefore, funding for mitigation should be borne by the state and federal governments and Imperial Valley farm owners (through a property assessment) to implement meaningful Salton Sea restoration. Concurrently, and second, immediate measures are necessary to restore wetlands in the Colorado River delta to benefit wildlife and the indigenous Cucapá, who endured a century of poverty resulting from America's destruction of the Colorado River estuary.

Next, state and federal laws must clearly define reasonable and beneficial use of water. With this standard in place, the Imperial Irrigation District will be legally compelled to modernize its distribution system rather than handing the bill to urban residents. Clear definitions of water's reasonable and beneficial use have the added benefit of implementing efficient irrigation techniques and technologies. Fourth, climate change and dwindling water resources have raised natural resource scarcity to the level of national security. Food production and water use must be reconceptualized for social stability, affordability, and broad economic well-being. A century ago, Congress enacted laws to promote a public policy agenda predicated on broad social distribution of federal land and water resources. Lawmakers must return to this vision by eliminating subsidies for large-scale farm investors and enacting a new regime of laws and policies promoting ecologically sustainable, smaller-scale, owner-operated farming buoyed with financial incentives and technical assistance from state and federal experts.

Fifth, irrigating half a million acres of desert is no longer sustainable in light of drought, global warming, and urban population growth. Over 10 percent of the county's farmland is idled to transfer water to cities. Not only is it unconscionable public policy to pay farmers not to grow food, it is not in the public interest. A smarter, cheaper, and more ethical way to address water reductions is to scientifically identify, condemn under eminent domain, and permanently retire low-yielding marginal ground. Landowners should receive payment for their property from water transfer funds rather than paying them not to farm. Place retired land under the authority of the Bureau of Land Management with a mandate to reestablish desert habitat to heal the ecological wounds of a century of industrial irrigated agriculture. Finally, dependence on foreign agricultural labor is directly responsible for Imperial County's elevated unemployment, dependency on social welfare, and poverty. The United States must end the valley's reliance on Mexican migrant day laborers to free up jobs for Americans.

This book opens with my recollection of floating down the Colorado River in an inner tube. Today, the hardest-working river in America staggers under an impossible burden. For over a hundred years, agribusiness refused to acknowledge that river and desert embrace in a symbiotic relationship. Instead, farmers used water, land, and labor as commodities to be exploited and discarded. My recommendations offer a hopeful alternative for the region, one predicated on a more sustainable, equitable, and ethical future than the previous century.

Notes

Introduction

1. *BN,* 11 December 1939.

2. Recent studies: Vollmann, *Imperial;* Round, *Impossible Land.*

3. International histories: Evans, *Bound in Twine;* Santiago, *Ecology of Oil;* Blackbourn, *Conquest of Nature;* Lines, *Taming Great South Land;* Teisch, *Engineering Nature;* Tyrrell, *True Gardens of the Gods.*

4. James, *Wonders of the Colorado Desert,* 527–31.

5. Chase, *California Desert Trails,* 275, 284, 286.

6. US Congress, House, *Protection and Development of Lower Colorado River Basin,* 1.

7. *BN,* 10 May 1921.

8. Diverse manifestations of power: Gutiérrez, *When Jesus Came;* Haas, *Conquests and Historical Identities.*

9. Sykes, *Colorado Delta,* 49–51, 66–76, chapter 8.

10. Ibid.; Hundley, *Dividing the Waters;* Hundley, *Water and the West;* Hundley, *Great Thirst;* Worster, *Rivers of Empire;* deBuys and Myers, *Salt Dreams.*

11. Kelly, *Cocopa Ethnography;* Williams, *Cocopah People.*

12. Capitalistic transformation of American West: Isenberg, *Mining California;* Morrissey, *Mental Territories;* Robbins, *Landscapes of Promise;* Cronon, *Nature's Metropolis;* Limerick, *Legacy of Conquest.*

13. Recent contested labor relations: Jameson, *All That Glitters;* Andrews, *Killing for Coal.* Mexicali Valley: Garduño, *La disputa.*

14. California industrial agriculture: McWilliams, *Factories in the Fields;* McWilliams, *California.* Overview: Walker, *Conquest of Bread;* Henke, *Cultivating Science;* Sackman, *Orange Empire;* Stoll, *Fruits of Natural Advantage;* Pisani, *From Family Farm to Agribusiness.* Profarmer perspective: Vaught, *Cultivating California.* Mexicali: Anguiano Téllez, *Agricultura y migración;* Sánchez Ramírez, *Crónica agrícola;* Herrera Carrillo, *Reconquista y colonizatión;* Walsh, *Building the Borderlands.*

15. *CC,* 8 December 1939.

16. Racial borderlands: Cadava, "Borderlands of Modernity"; Gutiérrez, "Migration, Emergent Ethnicity"; Benton-Cohen, *Borderline Americans;* Meeks, *Border Citizens;* Chang, *Pacific Connections;* Shah, *Stranger Intimacy;* Truett, *Fugitive Landscapes;* Calderón, *Mexican Coal Mining Labor;* Leonard, *Making Ethnic Choices;* Romero, *Chinese in Mexico;* Velázquez Morales, *Los inmigrantes chinos.*

17. *CC,* 2 August 1922.

Chapter 1

1. *BN,* 19 February 1927; 16 September 1930; 1 February 1935.

2. Sykes, *Colorado Delta,* 13–23; Howe and Hall, *Story of the First Decade,* 24–27; Farr, *History of Imperial County,* 3–4, 11, 13; Tout, *First Thirty Years,* 25–26.

3. C. R. Rockwood, "Born of Desert," in Tout, *First Thirty Years,* 29–37; Howe and Hall, *Story of the First Decade,* 39–45; Frisby, *Imperial Valley,* chapter 3.

4. Heffernan, *Personal Recollections.* Other investors: Frisby, *Imperial Valley,* 23; Schonfeld, "Early Development: Part I," 279–307.

5. Hendricks, "Guillermo Andrade"; Heffernan, *Personal Recollections,* 2–4; Howe and Hall, *Story of the First Decade,* 48–49. Contract between Andrade and California Development Company: Cory and Blake, *Imperial Valley,* 1251–52.

6. Cory and Blake, *Imperial Valley,* 1250; Tout, *First Thirty Years,* 30; Howe and Hall, *Story of the First Decade,* 127; Heffernan, *Personal Recollections,* 3–4; Dowd, *History of Imperial Irrigation District,* 15–16; Frisby, *Imperial Valley,* chapter 4.

7. Heffernan, *Personal Recollections,* 6; Frisby, *Imperial Valley,* 31–34; Cory and Blake, *Imperial Valley,* 1253–54; Imperial Land, *From Desert to Garden,* n.p., CL; Alexander, *Life of George Chaffey,* 281; deBuys and Myers, *Salt Dreams,* chapter 6. Infighting among CDC shareholders: Farr, *History of Imperial County,* 4–5.

8. Smythe, *Conquest of Arid America,* xiii. Smythe's vision did not include Asians, ibid., 149, 309. Anti-Asian sentiments underlying reclamation: Boime, "Beating Plowshares."

9. Smythe, *Conquest of Arid America,* 43. Smythe's biography: ibid., 261–67; Lee, "William E. Smythe and San Diego"; Lee, "William Ellsworth Smythe and Irrigation Movement"; Limerick, *Desert Passages,* chapter 4.

10. Heffernan, *Personal Recollections,* 6; Alexander, *Life of George Chaffey,* 284–85, 294; Cory and Blake, *Imperial Valley,* 1258–59.

11. Frisby, *Imperial Valley,* 37–38; Schonfeld, "Early Development: Part I," 288; Alexander, *Life of George Chaffey,* 284–87; Tout, *First Thirty Years,* 47–48; Howe and Hall, *Story of the First Decade,* 57–63.

12. Cory and Blake, *Imperial Valley,* 1259–60; Alexander, *Life of George Chaffey,* 287–90; Tout, *First Thirty Years,* 35–36.

13. Imperial Land, *In a Nutshell,* n.p., CL; Imperial Land, *From Desert to Garden,* n.p., CL.

14. James, *Wonders of the Colorado Desert,* 63–64; deBuys and Myers, *Salt Dreams,* chapter 6; Alexander, *Life of George Chaffey,* 290.

15. Alexander, *Life of George Chaffey,* 290; Worster, *Rivers of Empire,* 196; Tout, *First Thirty Years,* 78–79; Cory and Blake, *Imperial Valley,* 1331–32.

16. Tisdale, "Cocopah Identity," chapter 3; Kelly, *Cocopa Ethnography.* Case study of similar themes in Sonora: Evans, "Yaquis vs. Yanquis."

17. Alexander, *Life of George Chaffey,* 297–99; Heffernan, *Personal Recollections,* 12–13; Tout, *First Thirty Years,* 35–36, 50; deBuys and Myers, *Salt Dreams,* 80–82; Frisby, *Imperial Valley,* 51.

18. Soil survey: Tout, *First Thirty Years,* 96–97. Boundary survey: Cory and Blake, *Imperial Valley,* 1269–70. According to local lore, the 1903–04 surveyors finished their task in a Yuma cantina: Howe and Hall, *Story of the First Decade,* 147–48.

19. Cultural and ideological foundations of 1902 Reclamation Act: Pisani, *To Reclaim a Divided West,* chapter 8; Pisani, *From Family Farm to Agribusiness,* chapter 10.

20. deBuys and Myers, *Salt Dreams,* 11.

21. Pisani, *To Reclaim a Divided West,* chapter 8. Overview of Bureau of Reclamation: Pisani, *Water and American Government.*

22. Mexico's November 1901 protest: Hundley, "Politics of Reclamation," 301–2; Frisby, *Imperial Valley,* 48–50.

23. Schonfeld, "Early Development: Part I," 289. Of 385 filings in 1904, only 160 final proofs were recorded: Tout, *First Thirty Years,* 67. Discredited soil report: ibid., 96–97.

24. Tout, *First Thirty Years,* 98–99.

25. Ibid., 97–99; Hosmer, "Imperial Valley," 44; deBuys and Myers, *Salt Dreams,* 91–93.

26. Quote is italicized: Tout, *First Thirty Years,* 99. Rejection of bill: Schonfeld, "Early Development: Part I," 298–99; Cory and Blake, *Imperial Valley,* 1272–73; Tout, *First Thirty Years,* 97. Bureau of Reclamation's interest in developing the Yuma and Imperial Valleys: Pisani, *From Family Farm to Agribusiness,* chapter 10. Smythe's motives and ambitions: Lee, "William E. Smythe and San Diego," 10–24.

27. Howe and Hall, *Story of the First Decade,* 130–31.

28. Hundley, *Water and the West,* 26.

29. Cory and Blake, *Imperial Valley,* 1275–76; Howe and Hall, *Story of the First Decade,* 101.

30. Tout, *First Thirty Years,* 97.

31. Unprotected gate: Cory and Blake, *Imperial Valley,* 1287, 1290, 1292–402.

32. Woodbury, *Colorado Conquest,* 168–70; Cory and Blake, *Imperial Valley,* 1272–73. Meetings to purchase CDC: Holt, *Imperial Valley Scrapbook,* 88–136, CL. Paul Van Dimas: Hosmer, "Imperial Valley," 45–46. Antigovernment sentiment: Schonfeld, "The Early Development: Part I," 299–300, 303; Schonfeld, "Early Development: Part II," 402, 412; Holt, "How the Reclamation Service," 510–12; Holt, "Reclamation Service and the Imperial Valley," 70–75.

33. In September and October 1904, and in his 1904–05 annual report of January 1905, the secretary of the interior rejected CDC's offer of $3 million: Cory and Blake, *Imperial Valley,* 1273. Government's objections: *Decisions of the Department of the Interior and General Land Office,* 391–408.

34. Tout, *First Thirty Years,* 36.

35. Sperry, "When the Imperial Valley," 9–25; deBuys and Myers, *Salt Dreams,* 99–103.

36. Howe and Hall, *Story of the First Decade,* chapter 8. 1905–07 floods: Sykes, *Colorado Delta,* 112–17; Tout, *First Thirty Years,* chapters 7–8; Schonfeld, "Early Development: Part II," 395–415.

37. Kennan, *Salton Sea;* James, *Wonders of the Colorado Desert,* 63; Sykes, *Colorado Delta,* 112–17; Tout, *First Thirty Years,* chapters 7–8.

38. James, *Wonders of the Colorado Desert,* chapters 35–36. Mexican commission: Mitchell, "Salton Sea," 224–36. Díaz administration sent engineers García and Blanco-Peña to investigate: Thomas H. Silsbee to Cal.-Mex. Land and Cattle Co., 18 November 1905, folder Letters 1905, box 289, SL; Duryea, "Salton Sea Menace," 3–24.

39. Southern Pacific Railroad's bulletin ran summary of floods: *BN,* 8 July 1921; James, *Wonders of the Colorado Desert,* 64.

40. Native Indians: Cory and Blake, *Imperial Valley,* 1330–33; Heffernan, *Personal Recollections,* 13; Farr, *History of Imperial County,* 25–28; Howe and Hall, *Story of the First Decade,* 116.

41. Tout, *First Thirty Years,* 104–5. Rurales: Vanderwood, *Disorder and Progress.*

42. *BN,* 8 July 1921.

43. Letters between Harriman and Roosevelt: Kennan, *Salton Sea,* 76–82; Cory and Blake, *Imperial Valley,* 1358–62.

44. Kennan, *Salton Sea,* 84–90.

45. Levee breaks: Cory and Blake, *Imperial Valley,* 1418–21, 1434–37. Harry Cory: Tout, *First Thirty Years,* 101, 103, 106–7; *BN,* 8 July 1921. Altered landscape: Sykes, *Colorado Delta,* 117–18; Kennan, *Salton Sea,* 99. Cocapá: Tisdale, "Cocopah Identity," 112–14, 137–39.

46. Commission consisted of Andrés Aldasoro, sub-secretary of state and of the Department of Fomento, Colonization and Industry; Fernando Beltrán y Puga, chief Mexican commissioner of the International Boundary Commission and inspector of the irrigation work along the Colorado River; and Roberto Gayol, official engineer of Mexico City and author of a publication on colonization and irrigation: DOA to General Harrison Gray Otis, 10 September 1907, folder 1907, box 290, SL. Mexico's incipient river surveys: Kroeber, *Man, Land, and Water,* chapter 5. Hosting the commission, DOA to J. M. Packard, 24 September 1907, folder 1907, box 290, SL. Beltrán y Puga's October 1907 report: Samaniego López, "El control del río Colorado," 64.

47. F. B. [Fernando Beltrán y] Puga to D. O. Anderson, 2 December 1908, folder 1908, box 290, SL; OFB to F. B. Puga, 7 December 1908, folder 1908, box 290, SL. Meeting: Porfirio D[í]az, Mexico City to General Harrison Gray Otis, 22 February 1909, folder 1909, box 290, SL; map, DOA to Enrique de la Sierra, 22 June 1909, folder 1909, box 290, SL; Laguna Project map, Enrique de la Sierra to David O. Anderson, 26 June 1909, folder 1909; box 290, SL.

48. CDC debt: Dowd, *History of Imperial Irrigation District,* 41; Tout, *First Thirty Years,* 109–11. Debt to railroad: Cory and Blake, *Imperial Valley,* 1270. Receivership: Cory and Blake, *Imperial Valley,* 1433.

49. Tout, *First Thirty Years,* 109; Cory and Blake, *Imperial Valley,* 1432–33; Dowd, *History of Imperial Irrigation District,* 43–44, 49.

50. Ockerson, *Report on Lower Colorado River.*

51. Mexico's surveillance and silencing borderland critics: Harris and Sadler, *Secret War in El Paso;* Raat, *Revoltosos;* Smith, "Andrés G. García: Venustiano Carranza's Eyes, Ears, and Voice," 355–86. Repression of Yaquis and banishment to Yucatán: Evans, *Bound in Twine,* chapter 3. Overview: Hall and Coerver, *Revolution on the Border,* chapters 2–3, 9; Sandos, *Rebellion in the Borderlands,* chapters 1–2. Police and federal agents enforcing the neutrality law in Los Angeles: *NYT,* 20 September 1927.

52. Blaisdell, *Desert Revolution,* chapter 1. Description of Mexicali: *NYT,* 5 March 1911.

53. *NYT,* 30–31 January 1911; 13 February 1911. Vega's attack: ibid., 16–18 February 1911; 5 March 1911; 8–9 April 1911; 16–17 April 1911; Taylor, "Fighting the Patria," 1–8; Gerhard, "Socialist Invasion," 295–301; Blaisdell, *Desert Revolution,* chapters 2–3. Indigenous Mexican magonistas: Weber, "Keeping Community," especially 221, 227–28. Levees: Ockerson, *Report on Lower Colorado River,* 80.

54. Blaisdell, *Desert Revolution,* 74, chapter 9; Taylor, "Fighting the Patria," 8–15.

55. Cory and Blake, *Imperial Valley,* 1437–49; Kennan, *Salton Sea,* 102–4. In 1930, Congress reimbursed the railroad $1 million: deBuys and Myers, *Salt Dreams,* 120–21.

56. Tout, *First Thirty Years,* chapters 11–12; Dowd, *History of Imperial Irrigation District,* 48–52; *ECP,* "San Diego & Arizona Railway," 33–34, 36; Moeller, *Phil Swing,* 9–19; Hundley, *Water and the West,* chapter 2. Background: Pisani, "Irrigation District," 257–92.

57. Andrade suing Holabird, W. H. Holabird to O. F. Brant, 6 November 1911; folder 1911, box 287, SL. Holabird jail: W. Holabird to John D. Works, 19 March 1912, Letters to Works, Da–K Miscellany File, Holabird, W., box 3, MSS/C-B 423, Inventory of the John D. Works Papers, BL. Antagonism between receivers: Cory and Blake, *Imperial Valley,* 1335, 1434; Howe and Hall, *Story of the First Decade,* 135–42; Tout, *First Thirty Years,* 119.

58. Ward, "Twentieth-Century Ghosts," 359–85. Congressional efforts to purchase or trade for Baja California: *NYT,* 3–4 August 1912; 9 May 1915; 26 January 1919. US Army plan to protect the Alamo Canal "from bands of marauders": *BN,* 9 August 1920.

59. Dowd, *History of Imperial Irrigation District,* 88–89; Tout, *First Thirty Years,* 119; Hundley, "Politics of Reclamation," 308–10.

60. Tout, *First Thirty Years,* 113–14; Dowd, *History of Imperial Irrigation District,* 49.

61. Dowd, *History of Imperial Irrigation District,* 52; *ECP,* "San Diego & Arizona Railway," 38; Frisby, *Imperial Valley,* 72; Schonfeld, "Early Development: Part II," 420–22. CDC operated the Alamo Canal from 1900 to June 22, 1916. During that period, the Southern Pacific Railroad managed the CDC from June 20, 1905, to December 13, 1909; CDC fell under receivership from December 13, 1909, to February 8, 1916. Southern Pacific Railroad assumed management of CDC from February 8, 1916, to June 22, 1916; Imperial Irrigation District purchased it June 23, 1916. Adams, *Irrigation from Colorado River in Lower California,* 12–13. IID floated bonds in 1915, 1917, 1919, 1922: Burgess, *Imperial Irrigation District,* 16; Tout, *First Thirty Years,* 118. Reappraisal of railroad's relationship with IID: Orsi, *Sunset Limited,* 232–37.

62. Translation of Article 27, Mexico's legal authority to regulate La Compañía and set water rate and tariff: US Congress, House, *Report of the American Section of the International Water Commission United States and Mexico,* 140–58. Mexico's stream regulation: Kroeber, *Man, Land, and Water,* chapter 6.

63. de Vivanco, *Baja California al Dia,* 337–42, 346, 353.

64. Unequal expenditures: Dowd, *History of Imperial Irrigation District,* 73. Concern about local water officials: H. H. Clark to C. L. G[ó]mez, 28 July 1920, folder Letters 1920, box 291, SL; CLG to H. H. Clark, 5 August 1920, folder Letters 1920, box 291, SL. Example of CRLC's unwillingness to cooperate with request for copy of company's land titles and a list of its leases requested by engineer Ygnacio Hoyo López of General Agency of the Agrarian Directorate in Northern District of Lower California in Tijuana, an agency of the Agricultura and Fomento in Mexico City: H. H. Clark to C. L. G[ó]mez, 8 April 1921, folder 134 Ingo Hoyo López—Solicitad Datos 1921, box 292, SL; Secretaria de Agricultura y Fomento to Luis Curtis Gómez, no date, folder 134 Ing. Hoyo López—Solicitad Datos 1921, box 292, SL. For a month Hoyo López reiterated his request and the CRLC held him off, Ygnacio Hoyo López to Luis Curtis Gómez, 15 April 1921, folder 134 Ingo Hoyo López—Solicitad Datos 1921, box 292, SL; H. H. Clark to C. L. G[ó]mez, 22 April 1921, folder 134 Ingo Hoyo López—Solicitad Datos 1921, box 292, SL. The company's secretary wrote its lawyer: "As you will see, this communication demands a lot from us, and the company does not want to give Hoyo López what he is asking for," Secretary C. L. G[ó]mez to I. Pizarro Su[á]rez, 25 April 1921, folder 134 Ingo Hoyo López—Solicitad Datos 1921, box 292, SL. The CRLC delayed Hoyo López's requests, arguing the data Hoyo López wanted "would probably be used in making trouble for the company," Secretary C. L. G[ó]mez to Arturo Guajardo, 25 April 1921, folder 134 Ingo Hoyo López—Solicitad Datos 1921, box 292, SL; Arturo Guajardo to Luis Curtis Gómez, 26 April 1921, folder 134 Ing. Hoyo López—Solicitad Datos 1921, box 292, SL; Arturo Guajardo to C. L. Gómez, 28 April 1921, folder 134 Ing. Hoyo López—Solicitad Datos 1921, box 292, SL; CLG to Arturo Guajardo, 29 April 1921, folder 134 Ing. Hoyo López—Solicitad Datos 1921, box 292, SL. The company's lawyer informed the stockholders that private landowners did not have to furnish this information to the government, Pizarro Suárez to C. L. G[ó]mez, 4 May 1921, folder 134 Ingo Hoyo López—Solicitad Datos 1921, box 292, SL; Arturo Guajardo to C. L. Gómez, 20 May 1921, folder 134 Ing. Hoyo López—Solicitad Datos 1921, box 292,

SL; CLG to Arturo Guajardo, 20 May 1921, folder 134 Ing. Hoyo López—Solicitad Datos 1921, box 292, SL; Colorado River Land Company to Ignacio Hoyo Lopez, 20 May 1921, folder 134 Ing. Hoyo López—Solicitad Datos 1921, box 292, SL. Permit to draw water from Colorado River, Ismael Pizarro Suárez to Under-Secretary in Charge of the Department of Agriculture and Fomento, 1 August 1927 [English translation], folder 114 Projects Fraccionamiento Brant—1926, box 292, SL.

65. Tout, *First Thirty Years,* chapter 12; Dowd, *History of Imperial Irrigation District,* 49–93.

66. Adams, *Report of Inspection Trip over Colorado River Levee Systems below Yuma, Arizona,* 13–14. Changes in delta after 1909: US Congress, House, *Protection and Development of Lower Colorado River Basin,* 135–43.

67. Dowd, *History of Imperial Irrigation District,* 89–92; Hundley, *Water and the West,* 36–52.

68. US Congress, House, *All-American Canal in Imperial and Coachella Valleys;* Moeller, *Phil Swing,* 6–8; Hundley, "The Politics of Reclamation," 292–325.

69. Hundley, *Water and the West,* chapters 3–4, especially, 66–74.

70. US, Department of the Interior, *Preliminary Report on Problems of Imperial Valley and Vicinity;* Hundley, *Water and the West,* 8–16, chapters 3–5.

71. Dowd, *History of Imperial Irrigation District,* 65–68. 1922 bond election to buy the thirteen mutual water companies: Tout, *First Thirty Years,* 126–27.

72. US Congress, House, *Protection and Development of Lower Colorado River Basin,* 8–9.

73. *BN,* 23 June 1922.

74. Hundley, *Water and the West,* chapters 6–7; 337–43.

75. *BN,* 24 February 1927. Senate adjourned without voting on bill: ibid., 15 March 1928. Campaign for Boulder Canyon Act: Moeller, *Phil Swing,* chapters 2–7. Compact and passage of Boulder Canyon Act: Hundley, *Water and the West,* chapter 8.

76. *BN,* 15 March 1928.

77. US Congress, Senate, *Boulder Canyon Reclamation Project,* 19. Johnson's remark: ibid., speech 4 June 1924, following p. 88.

78. Johnson, *Epic of Imperial Valley,* n.p., CL.

79. Hundley, *Dividing the Waters,* 55; Boime, "National Moat," 164, 167. CRLC's argument against the bill: *Imperial Valley's Most Essential Need is a Flood Control and Storage Dam,* SL.

80. *BN,* 3 May 1923; IID's rebuttal, ibid., 5 May 1923.

81. Boime, "Beating Plowshares"; Boime, "National Moat." Japanese colonization proposals and opposition: *CC,* 20 March 1919; 1–3 April 1919; *BN,* 31 January 1920; 18 November 1922; 19 December 1922; 30 December 1922; 26 July 1924; 30 October 1924. *El Mercurio,* ibid., 23 October 1924.

82. Adams, *Progress Report of Special Colorado River Investigations,* 88–90. Additional evidence IID lacked adjudicated water rights: Frank Adams, Memorandum for Dr. [Elwood] Mead, Allocation of Colorado River Water to Mexico, April 11, 1929, p. 4–5, folder 304, box 20, FA.

83. *BN,* 21 December 1928; 22 December 1928. Text and evaluation of Boulder Canyon Act: Tout, *First Thirty Years,* 142–50. Passage of Boulder Canyon Act and Arizona's opposition: Hundley, *Water and the West,* chapters 8–9; Hundley, "Clio Nods."

84. Hundley, *Great Thirst,* 476–77.

85. *Answer to Brockman-Beal Letter,* n.p.

86. IID's intramural debates: *BN,* 24 October 1931; 2 December 1931; 8 December 1931; 12 December 1931; 19 December 1931. California legislative approval: ibid., 30 December 1931. Vote: ibid., 12 February 1932. Criticism of IID board: ibid., 2 April 1932; 5 April 1932; 8–9 April 1932; 14 April 1932; ibid., 24 June 1932.

87. *BN,* 2 December 1932. News articles explained contract: ibid., 3 December 1932; 5 January 1933. US Department of the Interior, *Boulder Canyon Project All-American Canal: Contract for Construction of Diversion Dam.*

88. Voters overwhelmingly approved canal funding: *BN,* 13 January 1933. Decline in farm profits: Dowd, *History of Imperial Irrigation District,* 77–83. IID taking mortgages: *BN,* 13 September 1933; Tout, *First Thirty Years,* 133.

89. Fradkin, *River No More,* 276–77. Circumvention of 160-acre restriction: Worster, *Rivers of Empire,* 156–88.

90. Hundley, *Great Thirst,* 222–23; Fradkin, *River No More,* 276–78.

91. Dissention IID's lobbying expenditures: *BN,* 25 March 1933; 26 April 1933. Swing's contract: ibid., 1 May 1933. Criticisms: ibid., 24 May 1933. Ickes's approval: ibid., 25 May 1933. Criticism of IID: ibid., 29 May 1933; 12 June 1933; 14–15 June 1933. Uproar over Phil Swing's consulting fees: Moeller, *Phil Swing,* 41, 70, 76–79; *BN,* 12 June 1925.

92. Fradkin, *River No More,* 276–278. Judge's ruling: *BN,* 5 July 1933.

93. Swing's lobbying: *BN,* 21 July 1933; 10 August 1933; 18 August 1933; 21 August 1933; 19 October 1933. Headline: ibid., 24 October 1933. Celebrations: ibid., 25–26 October 1933. Swing's speech summarizing canal politics: ibid., 3 November 1933.

94. Ibid., 23 June 1934; 2 July 1934. Swing retained to lobby during the water shortage: ibid., 3 July 1934; 21 July 1934. Canal work began: ibid., 8 August 1934. Imperial Dam: ibid., 12 July 1934. Canal issues: 26 July 1934; 11 August 1934. Debate over paying farmers to dig the canal rather than machines (called "force accounts"): ibid., 14 September 1934; 19 September 1934; 27 September 1934. Rose: ibid., 10 August 1934. Force accounts began: ibid., 22 October 1934. Mike Dowd reviewed All-American Canal campaign: ibid., 15 December 1934. Governor attending groundbreaking: ibid., 17 December 1934.

95. Hundley, *Great Thirst,* 223.

96. Hundley, *Dividing the Waters,* 134–35; Hundley, *Water and the West,* 298–99; Hundley, "The Politics of Water," 217–23.

97. Tout, *First Thirty Years,* 113.

98. US Department of Interior, Bureau of Reclamation, *Hoover Dam,* n.d., folder Herbert Hoover, box 333, HH.

Chapter 2

1. James, *Wonders of the Colorado Desert,* frontispiece.

2. Wright, *Winning of Barbara Worth,* 86.

3. Alexander, *Life of George Chaffey,* 285; Howe and Hall, *Story of the First Decade,* 63; Farr, *History of Imperial County,* 15; Imperial Land, *From Desert to Garden,* n.p.

4. Farr, *History of Imperial County,* 34, 95; Schonfeld, "Early Development: Part I," 289–91; Tout, *First Thirty Years,* 66, 89, 170, 172.

5. Imperial Land, *From Desert to Garden,* n.p. Imperial Land promotional literature: Holt, *Imperial Valley Scrapbook,* CL. Desert Land Act: Imperial Land, *In a Nutshell,* 45–47, CL. Western boosterism: Wrobel, *Promised Lands,* chapters 3–4; Morrissey, *Mental Territories.*

6. *Imperial Press,* 22 June 1901; Tout, *First Thirty Years,* 66–67, 87; Alexander, *Life of George Chaffey,* 278; Frisby, *Imperial Valley,* 24, 41–42.

7. Lavender, *California,* 320. CDC officers filed for land and sold it or donated it to start towns: Heffernan, *Personal Recollections,* 9. Water stock: Howe and Hall, *Story of the First Decade,* 76, 78, 98; *Imperial Valley, California,* 13.

8. Railroad's donation: *BN,* 8 July 1921. Railroad's colonizing assistance: Howe and Hall, *Story of the First Decade,* 94–95; Farr, *History of Imperial County,* 9; Tout, *First Thirty Years,* 65, 73, 77. A positive reassessment: Orsi, *Sunset Limited,* 226–37.

9. Hosmer, "Imperial Valley," 43; Tout, *First Thirty Years,* 56.

10. Stowe, "Pioneering Land Development," 15–39. C & M ranch: "Imperial Valley," *CC,* 2nd Annual Magazine, 61–62; Tout, *First Thirty Years,* 78. Sympathetic assessment of CRLC: Kerig, "Yankee Enclave." Critical assessment of CRLC: Herrera Carrillo, *Reconquista y colonización.*

11. Tout, *First Thirty Years,* 79. Chinese in Mexicali: Velázquez Morales, *Los inmigrantes chinos;* Gerardo, *El dragón.*

12. Howe and Hall, *Story of the First Decade,* 63; Cory and Blake, *Imperial Valley,* 1259; deBuys and Myers, *Salt Dreams,* 80–81.

13. Frisby, *Imperial Valley,* 37, 42–45. Land script: Alexander, *Life of George Chaffey,* 307; Robinson, *Land in California,* chapter 13, especially p. 177.

14. Lavender, *California,* 321.

15. *Imperial Valley, California,* 15.

16. Dowd, *History of Imperial Irrigation District,* 52; Frisby, *Imperial Valley,* 72; Schonfeld, "Early Development: Part II," 420–22.

17. *Thirteenth Census, Abstract of the Census, Supplement for California,* 668; ibid., *Fifteenth Census,* 1930, *Irrigation of Agricultural Lands,* 94.

18. Tout, *First Thirty Years,* 85–86.

19. Ibid., 57, 151–52.

20. "Imperial Valley," *CC,* 2nd Annual Magazine, 70, 72, 74; Farr, *History of Imperial County,* 17, 45–47, 287–90; Howe and Hall, *Story of the First Decade,* 160–61; Tout, *First Thirty Years,* 53–54, 71, 309; Hosmer, "Imperial Valley," 43.

21. Howe and Hall, *Story of the First Decade,* 160–61, 177–78; Tout, *First Thirty Years,* 53–54, 71, 309, 324–25; "Imperial Valley," *CC,* 2nd Annual Magazine, 70, 72, 74; Hosmer, "Imperial Valley," 43; Farr, *History of Imperial County,* 20, 22, 281–83.

22. Tout, *First Thirty Years,* 55–56.

23. Farr, *History of Imperial County,* 8, 34–35.

24. Quote: Pisani, "Irrigation District," 261. Overview: Pisani, *From Family Farm to Agribusiness.*

25. Wright, *Winning of Barbara Worth.* Intersection of capitalism and nature: Henderson, *California and the Fictions of Capital,* chapter 6.

26. Wright, *Winning of Barbara Worth,* 30, 154, 217–18.

27. Ibid., 54, 96.

28. Harold Bell Wright Gallery, PM; *The Winning of Barbara Worth,* produced by Samuel Goldwyn.

29. Tout, *First Thirty Years,* 177; Farr, *History of Imperial County,* 30; Frisby, *Imperial Valley,* 46–48, 53–54; Howe and Hall, *Story of the First Decade,* 77–80, 147–50, Cory and Blake, *Imperial Valley,* 1270.

30. Howe and Hall, *Story of the First Decade,* 197. District attorney Phil Swing indicted land sharpers gang in 1911: Schonfeld, "Early Development: Part II," 418–19. Vigilante justice against a land sharper: Tout, *First Thirty Years,* 312.

31. Howe and Hall, *Story of the First Decade,* 152.

32. Ibid., 154–55.

33. Ibid., 152–53; Hosmer, "Imperial Valley," 42–43.

34. *BN,* 8 March 1912.

35. Tait, *Irrigation in Imperial Valley.*

36. [Calipatria] *Herald,* anniversary edition, 31 March 1955; Tout, *First Thirty Years,* 348–49; Balfour-Guthrie: *BN,* 17 September 1920.

37. *Thirteenth Census, Abstract of the Census, Supplement for California,* 639; ibid., *Fifteenth Census,* 1930 vol. 2, pt. 3, *Agriculture,* 524.

38. *BN,* 13 January 1920.

39. Ibid., 10 April 1920; 20 September 1920. Quote: 12 July 1920.

40. Ibid., 14 October 1920.

41. Ibid., 1 April 1921; 7–8 April 1921; 10 May 1921; 13 May 1921; 21 June 1922. Under the Smith-Hughes Act, Veterans Bureau in charge of "trainee" program: ibid., 24 October 1921. Veteran's Bureau shipped "trainees" from coast to learn farming and ranching before they purchased land: ibid., 10–11 May 1921. By August 1922, several hundred trainees in valley: ibid., 10 August 1922.

42. Ibid., 24 November 1922.

43. Andrés, "Power and Control," table 1, 96. European immigrants: ibid., appendix 1, p. 519.

44. *Thirteenth Census, Abstract of the Census, Supplement for California,* 600–601; Andrés, "Power and Control," table 1, 96.

45. J. M. Davis, *Report of J. M. Davis to Imperial County, October 7 to 9, 1924,* folder 5, 18–5, Reports, California, Southern. Imperial County, box 18, Office File, 1914–1927, SRR, http://collections.stanford.edu/srr/bin/search/simple/process?query=imperial+county.

46. Andrés, "Power and Control," appendix 2, p. 520. Local ethnic organizations: Anderholt and Anderholt, *History of Imperial Valley Swiss;* Leonard, *Making Ethnic Choices.* Studies of rural California ethnic community building: Alamillo, *Making Lemonade;* Barajas, *Curious Unions;* García, *A World of Its Own.*

47. Aguirre Bernal, *Compendio histórico-biográfico de Mexicali,* chapter 3.

48. Cantú's coup: *NYT,* 14 December 1914. Laudatory overview of Cantú's career: Aguirre Bernal, *Compendio histórico-biográfico de Mexicali,* chapter 4. Cantú's support of agriculture and Chinese immigration: Gómez Estrada, *Gobierno y casinos,* 44–53. Anti-Chinese riots and Sinophobia: Romero,

Chinese in Mexico, chapter 6. Chinese transnational migration to northern Mexico: Romero, *Chinese in Mexico,* chapter 2. The Chinese in Mexicali: Duncan, "Chinese and Economic Development"; Cantú's bribes: [Curtis Gómez to] O. F. Brant, 26 March 1920, folder Letters 1920, box 291, SL. Complaints about tax collectors and taxes: [Curtis G[ó]mez to] O. F. Brant, 22 May 1918, folder Letters 1918, box 291, SL.

49. Cantú's vice policies: Gómez Estrada, *Gobierno y casinos,* 43–55; Vanderwood, *Satan's Playground;* 83; Recio, "Drugs and Alcohol"; Sandos, "Northern Separatism"; Schantz, "All Night at the Owl." Comparative analysis: Moore, "Refugees from Volstead." Mexicali population statistic: Duncan, "Chinese and Economic Development." Census by nationality: Chamberlin, "Mexican Colonization," 48. Cantú's reign: *NYT,* 22 August 1920; Kenamore, "Principality of Cantu," 23–28; Carr, "Kingdom of Cantu," 33–34, 65–67. Ouster: *NYT,* 29–31 July 1920; 3 August 1920; 8 August 1920; Aguirre Bernal, *Compendio histórico-biográfico de Mexicali,* 200–205.

50. In the 1920 census, 2.7 percent California's Mexican "alien" men over age twenty-one were naturalized US citizens: *Mexicans in California,* 63.

51. Example of marginalized working-class whites: *BN,* 20 February 1920.

52. Tout, *First Thirty Years,* 260; Howe and Hall, *Story of the First Decade,* 165.

53. *BN,* 29 March 1912. Quote: ibid., 21 June 1912. Imperial's marshal: ibid., 29 October 1914. Brawley's police commissioner: ibid., 17 March 1915.

54. Sources, in order: ibid., 4 May 1914; 4 April 1919; 5 February 1921. Quotes: ibid., 16 May 1924. More examples: ibid., 31 March 1924; 13 June 1924. Expulsions and deportations: ibid., 19 June 1929.

55. Vanderwood, *Satan's Playground.* 1924 border closure controversy: St. John, *Line in the Sand,* 166–72.

56. Richard Laughrin interview; Takanori (Pro) Nimura interview, 19 March 1998.

57. *BN,* 2 January 1912; 22 March 1912. "Negro quarter": *SD,* 5 April 1992. Overview of property ownership and residential segregation: Taylor, *Mexican Labor,* 65–71, 79–94. El Centro's segregated commercial districts and racism: Delia Gongora interview, 24 November 1999.

58. *BN,* 22 March 1912.

59. Ibid., 5 April 1927; Holliday, "A Study of Non-Attendance," 8–9. Quote: Taylor, *Mexican Labor,* 81–82. Overview of ethnic Mexicans: Ruiz, *From Out of the Shadows,* chapters 1–4.

60. "Interview with A. W. Swanson, El Centro," folder 332, "A Survey of Race Relations on the Pacific Coast," p. 1, box 36, SRR. Race relations: [Nora E. Kreps], *Diary of a Schoolteacher in the Imperial Valley, April 16, 1924,* folder 57, "A Survey of Race Relations on the Pacific Coast," 1925, box 24, SRR; Cramp, Shields, and Thomson, *Study of the Mexican Population;* Andrés, "I Am Almost More at Home."

61. *BN,* 20 August 1912. Quote: ibid., 3 June 1913.

62. Ibid., 30 April 1912.

63. Ibid., 21 September 1928; 26 September 1921.

64. Quote: ibid., 3 May 1922. Ibid., 22 September 1922.

65. Imperial Valley Snapshots, Ca. 1903, BANC PIC 19xx.054—PIC, Group 1, item numbers 1–2, BL.

66. James, *Wonders of the Colorado Desert,* 413.

67. *BN,* 28 July 1920.

68. Ibid., 5 September 1923; 13 September 1923; 15 September 1923.

69. Ibid., 10 September 1923; 17 July 1939.

70. Transformation of wilderness: Fiege, *Irrigated Eden,* chapters 1–2; Knobloch, *Culture of Wilderness;* Wyckoff, *Creating Colorado.*

71. Modesto Ibarra interview; Gerrard: *BN,* 22 September 1931.

72. Modesto Ibarra interview; Richard Laughrin interview.

73. *BN,* 19 April 1923. Allen: ibid., 7 January 1925.

74. Major landowners in Mexicali Valley, acreage statistics, and comparison of water ordered on both sides of border: Adams, *Irrigation from Colorado River in Lower California,* 6, 13, 28–29. Torres-Martínez reservation: deBuys and Myers, *Salt Dreams,* chapter 9. Common knowledge IID massively overused water to maintain a historical record of usage: McWilliams, *California,* 301–8.

75. Dowd, *History of Imperial Irrigation District,* 69–72, 76. Field tile drains: Burgess, *Imperial Irrigation District.*

76. US Department of Agriculture sending plant expert for CRLC, unsigned memo, 19 September 1907, folder 1907, box 290, SL; OFB to C. V. Piper, 17 December 1907, folder 1907, box 290, SL. Field station: *BN,* 12 February 1920. Seed and plant inspection for farmers on both sides of the border: Bell and Mackenzie, *Mexican West Coast,* 304.

77. Cooper, *Aqueduct Empire,* 75–76. Tile-laying machine demonstration: *BN,* 2 June 1933.

78. Report revealed IID furnished water "at a nominal cost to any farmer who uses it for purposes of reclaiming alkali land": *BN,* 6 March 1925.

79. Dowd, *History of Imperial Irrigation District,* 75; Fradkin, *River No More,* 182.

80. *BN,* 1 October 1925; *Thirteenth Census, Abstract of the Census, Supplement for California,* 668; *Fifteenth Census,* 1930, *Irrigation of Agricultural Lands,* 94. Desilting: Nadeau, *Water Seekers,* 229–33; Fradkin, *River No More,* 278–79.

81. Howe and Hall, *Story of the First Decade,* 167–68. Weed infestation ordinance: *BN,* 10 July 1923. Weed control ideology: Knobloch, *Culture of Wilderness,* chapter 4. Harrigan: *BN,* 5 June 1929. Equation of animal "pest" control with war: McNaught, "Squirrel Is a Hun"; "War on Aphids," *CC,* 16 April 1929; "Arms for Battle," ibid., 19 February 1935; "War on Coyotes," ibid., 9 December 1937. Studies on this topic: Russell, *War and Nature;* McWilliams, *American Pests.*

82. *BN,* 18 February 1926; 17 September 1938.

83. Ibid., 6 August 1935.

84. Ibid., 15 September 1925.

85. Ibid., 15 May 1926; 28 May 1926. Eliminating mildew: ibid., 25 January 1927.

86. Ibid., 1 October 1928.

87. Ibid., 9 August 1933; 24 July 1937; 4 May 1938; 12 May 1938. Three hundred men on public assistance spraying poison to kill grasshoppers: ibid., 17 April 1934; 21 April 1934. 1939 "hopper" campaign: ibid., 5 May 1939; 8 May 1939.

88. Evelyn Zinn interview; *BN,* 7 January 1937. Pest eradication: Walker, *Conquest of Bread,* 185–93.

89. *BN,* 23 September 1921. Letter to senator: ibid., 3 March 1924.

90. Ibid., 12 October 1921; 17 February 1926. Another article: *CC,* 23 April 1925.

91. *BN,* 6 July 1925; 24–25 July 1925. Scientific gopher eradication: Dixon, "Rodents and Reclamation"; Dixon, *Control of the Pocket Gopher.* Early references to gopher bounty: *CC,* 5 July 1919; 7 April 1921; 22 April 1921.

92. *BN,* 24 December 1925; 4 February 1926; 2 March 1926.

93. Quote: ibid., 1 February 1926. Additional references to gopher extermination: *CC,* 6 January 1927; 29 January 1927; 8 February 1927; 10 February 1927.

94. *CC,* 15 October 1928.

95. *BN,* 3 April 1929. Rabbit hunt: ibid., 3 May 1929; 18 January 1930.

96. Fiege, *Irrigated Eden,* 174.

97. B. A. Harrigan, ed. [untitled pamphlet, circa 1925], folder 21, carton 10, MSS 84/38C, PT. Contemporary pursuit of crop perfection, the quest to engineer the ideal carrot: *IVP,* 26 February 2014.

98. *BN,* 14 May 1926, section 2.

99. Ibid., 10 May 1921.

Chapter 3

1. Walker, *Conquest of Bread.*

2. Azuma, *Between Two Empires;* Roediger, *Wages of Whiteness.*

3. State Board of Control, *California and the Oriental.* Background on nativism, immigration restrictions, and racial farming in California: Street, *Beasts of the Field;* Ichioka, *Issei;* Leonard, *Making Ethnic Choices;* Spickard, *Almost All Aliens,* chapters 5–7; Ngai, *Impossible Subjects,* chapter 3.

4. Barnett, *Imperial Valley.*

5. "Mexican Labor in U.S. Field Notes, 16 October 1928," folder 10, carton 10, Series A, Set I, MSS 84/38C, PT; Anderholt, *Desert Dairies.*

6. Sano, *One Thousand Days,* 10–11; Takanori (Pro) Nimura interview, 18 March 1998.

7. *Thirteenth Census, Abstract of the Census, Supplement for California,* 638–39, 644–45. Irrigated acreage: Bureau of Census, *Statistical Atlas,* 1914, plate 391.

8. *Thirteenth Census, Abstract of the Census, Supplement for California,* 638, 650–51, 657. In 1909, Japanese growers grew cantaloupes and dominated the crop by 1912, folder 10 [handwritten notes], carton 10, MSS 84/38C, PT.

9. *BN,* 8 May 1926; also 5 May 1926. Farming and credit: Anderholt and Anderholt, *History of Imperial Valley Swiss,* chapters 1–3. All-American Canal bond hearing: US Congress, House, *All-American Canal in Imperial and Coachella Valleys.* Agricultural credit: Henderson, *California and the Fictions of Capital,* chapter 2.

10. Distributors and credit: Rush, "Imperial Valley, the Wonderful," 67–68. Denial of federal loans 1919–1941, loans available when All-American Canal went operational: Burgess, *Imperial Irrigation District,* 27.

11. "A Survey of Race Relations on the Pacific Coast," 1925, p. 1, folder 84, box 25, SRR.

12. Earl Fruit: Chloe Holt, 27 February 1925, p. 2, 5, folder 297, box 31, SRR. Dr. E. E. Chandler, "A Survey of Race Relations on the Pacific Coast," 1925, p. 2, folder 84, box 25, SRR.

13. Dr. E. E. Chandler, "A Survey of Race Relations on the Pacific Coast," 1925, p. 1, folder 84, box 25, SRR.

14. Taylor, *Mexican Labor,* 6–7, 69–71; Duke, "California the Golden," 245. Meaning of "white trash": Foley, *White Scourge.* Chinese "faithful compliance" and "first-class credit risks": Bell and Mackenzie, *Mexican West Coast,* 295, 311.

15. Stereotype Mexicans: Reisler, "Always the Laborer," 231–54. Quote: Chloe Holt, 27 February 1925, p. 7, folder 297, box 31, SRR.

16. Bell and Mackenzie, *Mexican West Coast,* 164, 295.

17. Taylor, *Mexican Labor,* 3–33. Farm acreage: ibid., 9. Film: *BN,* 18 November 1915. Fifteen-mile cotton ranch: Duke, "California the Golden," 244. Emulating CRLC: *BN,* 25 October 1910.

18. Emil Engeleke to J. H. Packard, 9 November 1905, folder Letters 1905, box 289, SL. Imaginary line quote: OFB to C.M. Davis, 22 July 1908, folder 1908, box 290, SL. 1908 quote: OFB/AAM to W. K. Bowker, 14 March 1908, folder 1908, box 287, SL. Eighty percent: Bell and Mackenzie, *Mexican West Coast,* 305. Study of mono-crop agribusiness: Evans, *Bound in Twine.*

19. Hu-DeHart, "Chinese of Baja California," 13. Chinese cooperative farming and lending: de Vivanco, *Baja California,* 198; Bell and Mackenzie, *Mexican West Coast,* 294, 305–6. Eight Chinese cotton tenants needing loans of $50,000–$100,000 within 30–60 days to pay for provisions, water, and picking costs: C. B. Williams to Harry Chandler, 20 July 1920, folder Letters 1920, box 291, SL. In 1921, CRLC began lending directly to tenants because lenders charged 25–100 percent interest per annum to CRLC's tenants. This outrage meant tenants were not making money and often were not able to pay the lease, forcing the company to lend money to tenants and banks to finance the tenants. CLG to Victor Carusso, 30 November 1921, folder Letters 1921, box 291, SL.

20. *Agrarista* demands for ejidos: Boyer, *Becoming Campesinos;* Bantjes, *As If Jesus Walked on Earth;* Craig, *First Agraristas,* chapter 3; Sanderson, *Agrarian Populism,* chapter 4; Hall, "Alvaro Obregón"; Hart, *Empire and Revolution,* 11.

21. Ismael Pizarro Suárez to Harry Chandler, 1 August 1927, folder 114 Projects Fraccionamiento Brant—1926, box 292, SL; Ismael Pizarro Suárez to the Under-Secretary in Charge of the Department of Agriculture and Fomento, 1 August 1927 [English translation], folder 114 Projects Fraccionamiento Brant—1926, box 292, SL.

22. *NYT,* 22 April 1928. CRLC's farming operation: Kerig, "Yankee Enclave." Description of Mexicali's cotton industry: de Vivanco, *Baja California,* 202, 204, 361–83, 391–95. In 1921, CRLC was so dissatisfied with poor ginning and exceedingly high ginning prices, the company changed long-standing policy and invested $125,000 to improve existing gins or build new ones: CLG to Victor Carusso, 30 November 1921, folder Letters 1921, box 291, SL.

23. Japan Cotton Trading Company: *BN,* 9 January 1920; 15 April 1921. Japan buying most of the cotton: Bell and Mackenzie, *Mexican West Coast,* 304. Earlier food buyers from Japan suggested planting rice in Imperial Valley: *BN,* 28 September 1914. Quote black farmers: *SD,* 5 May 1992.

24. *BN,* 28 March 1924. Early cantaloupe production, folder 10, carton 10, MSS 84/38C, PT. 12,195 railroad cars: Imperial County, California, *Annual Agricultural Crop and Livestock Report, 1928.* Com-

prehensive accounting of cantaloupe expenses: *BN,* 5 May 1920. W. S. Fawcett, a substantial grower and shipper, recalled the development of cantaloupe industry: "A Survey of Race Relations on the Pacific Coast," 1925, pp. 1–2, folder 72, box 25, SRR.

25. *Thirteenth Census, Abstract of the Census, Supplement for California,* 638, 644; *Fifteenth Census,* 1930 vol. 2, pt. 3, *Agriculture,* 524; Taylor, *Mexican Labor,* 9; US Congress, Senate, *Violations of Free Speech and Rights of Labor,* Hearings Pursuant to Senate Res. 266 76th Cong., Washington, DC: Government Printing Office, 1940, Exhibit 8900, pp. 20285–86, Exhibit 8901, p. 20286.

26. Official from Australia: *CC,* 3 September 1920.

27. *BN,* 5 May 1920. S. Kinoshita: ibid., 12 June 1922.

28. William Thornburg interview; IID 1927 crop and acreage report, "3–15–1928," folder 20, carton 10, MSS 84/38C, PT; IID 1938 crops by acreage report, Exhibit 8899, pp. 20284–85, *Violations of Free Speech.* The most productive crops in 1929 were hay (117,566 acres), cotton (22,165 acres), wide variety of vegetables (72,000 acres), lucrative perishable crops such as cantaloupes and other melons (31,000 acres), lettuce (25,000 acres), and peas (6,400 acres). *Fifteenth Census,* 1930 vol. 2, pt. 3, *Agriculture,* 538–39, 544, 552, 557, 562. In 1929, 1,378 farms reported owning 67,000 cows, fifty-eight ranches owned close to 56,000 sheep, 600 farms had over 33,000 turkeys. Ibid., 569.

29. Leonard, *Making Ethnic Choices,* chapter 2. Recent study: Chang, *Pacific Connections.*

30. Taylor, *Mexican Labor,* 6. Quote: *BN,* 2 December 1918. "Hindu menace": *BN,* 25 November 1918; 5 December 1918; Leonard, *Making Ethnic Choices,* chapters 2–3. Reappraisal South Asian homosexuality and race-sex policing: Shah, *Stranger Intimacy.*

31. E. E. Chandler, "A Survey of Race Relations on the Pacific Coast," 1925, p. 2, folder 84, box 25, SRR. White brother: *BN,* 20 May 1919.

32. Ichioka, *Issei,* chapter 6. T. Takimoto, *The California Alien Land [L]aw and Its Effect,* p. 5, folder, Miscellaneous Papers, box 6, SRR; Leonard, *Making Ethnic Choices,* chapter 3; *BN,* 3 July 1917. Circumventing 1913 Alien Land Law: ibid., 9 September 1921.

33. *BN,* 23 January 1918.

34. Ibid., 24 March 1919. Rally: ibid., 3 March 1919; *Fifteenth Census,* 1930 vol. 2, pt. 3, *Agriculture,* 519.

35. Ray Chikahisa interview.

36. *Tentative Findings of the Survey of Race Relations: A Canadian-American Study of the Oriental on the Pacific Coast, 1925,* pp. 16–17, folder Printed Material, box 1, SRR. Absentee landlord: "A Survey of Race Relations on the Pacific Coast," 1925, p. 1, folder 84, box 25, SRR.

37. Leonard, *Making Ethnic Choices;* Peter Iwao Sano interview; Takanori (Pro) Nimura interview, 18 March 1998; Ray Chikahisa interview.

38. *BN,* 17 October 1931. Putting farm titles in children's names: Leonard, *Making Ethnic Choices,* 56–57, 106–14, 135, 154.

39. Her name was Giraduo or Giruado: *BN,* 26 May 1928; 29 May 1928; 11 June 1928; 22 May 1931.

40. Ibid., 6 August 1936.

41. Ibid., 6 May 1919; 8 April 1936; 22 June 1936; 12 August 1939; 7 July 1939. A panoramic photo in the Filipino Gallery at Pioneers Museum, Imperial County, California, shows a picnic in 1939, probably celebrating the creation of the Filipino Grower Association. Recent studies of Filipinos: España-Maram, *Creating Masculinity;* Baldoz, *Third Asiatic Invasion.*

42. For elite Mexican opinions of Mexican farmers: Walsh, *Building the Borderlands,* chapter 3; Rodríguez, *Memoria administrativa,* 135–37. Stereotypes of Mexican "peons": González, *Culture of Empire.*

43. Guerin-Gonzales, *Mexican Workers,* chapter 2.

44. Sánchez, *Becoming Mexican American;* Andrés, "Invisible Borders."

45. Mexican bank accounts in Imperial Valley and remittances to Mexico, "21 June 1927 and 31 May 1927," folder 20, carton 10, Series A, Set I, MSS 84/38C, PT.

46. Taylor, *Mexican Labor,* 69, 71.

47. *BN,* 18 January 1917; 15 March 1920. Ramos: ibid., 29 October 1921.

48. Ibid., 9 April 1930; 8 August 1936; 30 July 1937; 8 August 1936; 12 January 1939.

49. Ibid., 30 November 1939; 11 January 1938. Exhibits 8905–8908, pp. 20291–94, *Violations of Free Speech; BN,* 22 January 1938.

50. J. M. Davis, *Report of J. M. Davis to Imperial County, October 7 to 9, 1924,* folder 5, 18–5,

pp. 1–5, Reports, California, Southern. Imperial County, Office File, 1914–1927, box 18, SRR (hereafter *Report of J. M. Davis to Imperial County*), http://collections.stanford.edu/srr/bin/search/simple/process?query=imperial+county.

51. *BN,* 21 April 1919. For tenant lifestyle: Peter Iwao Sano interview; *IVP,* Agricultural Edition, 30 April 2007.

52. *Thirteenth Census, Abstract of the Census, Supplement for California,* 638, 644; Exhibit 8900, pp. 20285–86, *Violations of Free Speech; Census of Agriculture,* 1945 vol. 1, pt. 33, *California: Statistics for Counties,* 88; R. L. Adams, *Commonwealth Club Tenancy Studies—April–June 1922,* p. 9, folder 277, box 30, SRR.

53. *Report of J. M. Davis to Imperial County,* 2, 4.

54. Ibid., 2.

55. *BN,* 27 October 1913; 26 January 1917.

56. Ibid., 12 January 1920. Tricolor map: ibid., 10 March 1920. Bill forbade Japanese ownership or leasing of farmland, forbade parents to be trustees or guardians of children's farmland, forbade Japanese corporations from leasing farmland. Punishment was a two-year prison sentence and/or a $500 fine. Ibid., 28 October 1920.

57. Ibid., 14 June 1920.

58. Ibid., 28 September 1920.

59. Ibid., 21 October 1920. Quote: 22 October 1920. See also ibid., 27 October 1920. Election results: ibid., 5 November 1920. A different tally for Imperial County (4,814 voted for the measure, while 1,996 cast ballots against): *Study of Vote Cast in 1920 on Alien Land Law,* p. 2, folder 415, n.d., box 37, SRR.

60. *BN,* 22 November 1920; 1 December 1920.

61. "A Survey of Race Relations on the Pacific Coast," 1925, p. 3, folder 93, box 25, SRR; Earl Fruit, Chloe Holt, 27 February 1925, p. 3, folder 297, box 31, SRR.

62. *BN,* 21 December 1921.

63. Yano case was a loophole for Asians to hold land in their children's names as guardians: ibid., 5 June 1922. For denying citizenship: López, *White by Law.* For 1920 Alien Land Law and other discriminatory laws and acts: *BN,* 20 February 1923.

64. *BN,* 19 November 1923; ibid., 13 December 1923; ibid., 20 December 1923.

65. Ibid., 5 January 1924. I. V. Farm Lands Association leased nearly 4,000 acres around Calipatria, and it replaced South Asians with white tenants: ibid., 18 January 1924; Alvin Jack, Imperial Valley—May 1934, pp. 5–6, folder 30, carton 14, MSS 84/38C, PT; *BN,* 5 January 1924. Blacks: ibid., 3 March 1924.

66. *BN,* 1 March 1924; 3 March 1924; 5–6 March 1924; 17 March 1924; 21 May 1924; 8 August 1924; 24 September 1924; 19 November 1924; 18 May 1925; 14 November 1925.

67. Ibid., 24 April 1924; 7 May 1924; 9 May 1924; 28 May 1924. An agriculturalist revealed Judge Conkling ruled against the law because his tenant was Japanese: "A Survey of Race Relations on the Pacific Coast," 1925, p. 3, folder 93, box 25, SRR. Appeal of Conkling's decision: *BN,* 24 July 1924.

68. *BN,* 27 December 1924; 12 January 1925; 18 November 1925. In 1934, the Supreme Court ruled section 9-A of the Alien Land Law unconstitutional because it made the defendants prove their innocence: ibid., 3 November 1934; Saund, *Congressman from India,* chapters 3–4.

69. *BN,* 19 December 1922; 30 December 1922; 26 July 1924; 23 October 1924. Japanese consul's visit to explore settlement opportunities in Baja California: ibid., 7 May 1924; 9 May 1924.

70. Ibid., 9 July 1925.

71. Ibid., 27 March 1926.

72. Ibid., 26 May 1931; 27 May 1931.

73. Ibid., 17 November 1920; 7 November 1923. Large growers dispensed with Farm Bureau: "Imperial Valley—May 1934," folder 302, carton 14, MSS 84/38C, PT; Exhibits 8905–06, pp. 20291–92, *Violations of Free Speech.* Economist Arthur Stuart's testimony: ibid., pp. 20126–35 and Exhibits 8,890–8,913, pp. 20284–97.

74. *BN,* 6 July 1929; 28 July 1930. Gerrard: ibid., 31 May 1929; 12 May 1931; 29 December 1932; 3 January 1933. Gerrard's end of receivership: ibid., 16 June 1934.

75. Exhibit 8903, p. 20130, *Violations of Free Speech.*

76. *Sixteenth Census,* 1940 vol. 2, pt. 3, *Agriculture,* 287. A different census calculated 2,116 farms

harvesting crops in 1939, totaling 274,808 acres: *Census of Agriculture,* 1945, 21. The acreage data was skewed because two crops could be grown in the same field in one year: Stuart, 20126–27, Exhibit 8899, pp. 20284–85, *Violations of Free Speech.* 1934 financial records: ibid., J. L. Leonard, Will J. French, and Simon J. Lubin, United States Special Commission on Agriculture Disturbances in Imperial Valley, *Report to the National Labor Board, February 11, 1934,* p. 20046 (hereafter Leonard Report), *Violations of Free Speech.* Standlee: CERA Field Notes—Clark Kerr, 5–22–1937, carton 14, folder 41, MSS 84/38C, PT. Desert farming difficulties: William Thornburg interview.

77. *Violations of Free Speech,* 20045. Quote: Imperial Valley—May 1934, p. 6, folder 30, carton 14, MSS 84/38C, PT.

78. *BN,* 14 September 1929.

79. Ibid., 30 March 1933. Government's involvement in farm marketing and California agribusiness violence against opponents: Woeste, *Farmer's Benevolent Trust.*

80. Ethnic diversity of growers and dairymen: Anderholt and Anderholt, *History of Imperial Valley Swiss.* By 1920, there were South Asian, Filipino, Japanese, and white grower associations in the county: *BN,* 6 May 1919. Fuller: Howe and Hall, *Story of the First Decade,* 199.

81. Farr, *History of Imperial County,* 187, 194–208. Farm Bureau statewide: Chambers, *California Farm Organizations,* chapter 3.

82. Yoshio Ichikawa, *Cooperative Movement among Japanese Farmers in California,* March 1925, pp. 10–11, 14–19, folder 333, box 32, SRR. List of interracial membership and acreage of Strawberry Growers Association: *BN,* 15 October 1923.

83. J. Y. Shiozawa, secretary of the Japanese Farmers Association, boasted cooperation: *BN,* 3 July 1917. Japanese farmers and shippers: ibid., 15 October 1923; 27 April 1921; Ichikawa, *Cooperative Movement,* 14–17.

84. McLachlan, *Community Production of Durango Cotton,* 1–16; Howe, "Calexico: Imperial Valley," 8–14; Kelly, "Calexico: King Cotton's Capital," 1–7. Chandler's organizing efforts: *CC,* 17 December 1919; 29 December 1919; 3 January 1920; 23 January 1920.

85. Mexican laborers imported to pick cotton: *CC,* 26 April 1920. Imperial Valley crop associations: *BN,* 12 February 1920; 23 September 1921, 4 September 1923. Clark president Chamber of Commerce: *CC,* 24 January 1923. Clark urged crop industry cooperation: ibid., 28 June 1923. Clark spoke to I. V. Farm Bureau: ibid., 3 January 1924. Clark elected president cotton association: ibid., 17 March 1924. Reelected: ibid., 11 March 1925.

86. Chester Moore: *BN,* 7 April 1926. 1925: Peters, "Agricultural Unionism," 116. Western Growers: *BN,* 29 March 1934.

87. *BN,* 5 February 1927; 10 February 1927; 11 February 1927. Movement toward a clearinghouse: ibid., 3 May 1928.

88. Ibid., 22 May 1931; 18 July 1931. Despite the melon association failure, growers formed Hay Growers' Association: ibid., 21 July 1938. Desert Grapefruit League: ibid., 13 March 1937, 6 April 1937. Beet Growers Association: ibid., 28 July 1937; 31 July 1937; 26 August 1937; 11 August 1938.

89. Exposé of Associated Farmers: Hosmer, "Who Are the Associated Farmers?" Hugh Osborne's creation and involvement in Associated Farmers: *BN,* 21 March 1934; 23 March 1934; 14 December 1934; ibid., 31 May 1939; 8 November 1939; 28 November 1939; 11 December 1939. Associated Farmers in Imperial County: Peters, "Agricultural Unionism," 132–46. As former commander of El Centro American Legion post, Osborne was named chairman of a special Americanism and anti-Red Commission of the American Legion: *BN,* August 17, 1934. For his announcement of two legislative proposals, one to stamp out subversive activities in California and another to build a statewide network to coordinate efforts to crush subversive elements: ibid., 4 January 1935. Osborne chaired the meeting to create chapters of Farmers' Protective Association (FPA) to oppose closed shop: ibid., 5 February 1938.

90. Summary of integrated agricultural cooperation: ibid., 9 November 1921.

91. Local history of cooperative marketing: ibid., 28 March 1924. How Bureau of Markets operated: ibid., 12 September 1921. Court ruling: ibid., 19 January 1922.

92. Ibid., 18 May 1922; 22 May 1922; 2nd edition, 25 May 1922; 28 June 1922; 21 October 1922.

93. Market Bureau renamed the Bureau of Agricultural Economics: ibid., 10 January 1923; 13 January 1923; 21 May 1924. For lettuce: ibid., 26 October 1925.

94. Ibid., special edition, 30 January 1922; 11 December 1925; 8 June 1922; 21 November 1922.

95. Quote: Ibid., 20 January 1926. Employment office: ibid., 20 November 1926; 22 November 1926.

96. Ibid., 28 September 1939. The Board of Supervisors budgeted $2,000 to an experiment station in addition to state funding: ibid. The Agricultural Extension Service, in 1923 announced a two-day soil fertility convention involving presentations from four University of California experts in agronomy, citriculture, and soil, as well as field demonstrations, lectures, and inspections. At night speakers gave two lectures: ibid., 5 March 1923.

97. Ibid., 6 July 1933; 8 July 1933. Forty-two farmers with 1,888 acres agreed to plow under 678 acres of crops in exchange for $1,538.50: ibid., 10 July 1933. For synthesis of New Deal and AAA's role in it: Kennedy, *Freedom from Fear,* chapter 7. Positive overview of New Deal ideology: especially chapter 6 (for water), Lowitt, *New Deal and the West.*

98. For why wheat growers rejected AAA's plan: *BN,* 17 July 1933; 26 July 1933; 5 August 1933. Imperial County farm subsidies: *Violations of Free Speech,* 20045, Exhibits 8909–9012, pp. 20295–96. Although 1,452 landowners received checks totaling $220,608, large owners received the biggest percentage: ibid., Exhibits 8909, pp. 20295, 20133. Farm Bureau urged members to sign worksheets for farm conservation program: ibid., 10 June 1937; 13 July 1936; 27 July 1936. "Soil Conserving" and "Soil Building" programs: ibid., 18 June 1936; 22 June 1936.

99. *BN,* 16 October 1937.

100. Ibid., 22 June 1936. Lobbying: ibid., 11 May 1938; 9 February 1939.

101. Ibid., 16 February 1938; 19 February 1938. Sugar beet subsidies: ibid., 9 November 1938; 2 December 1938; 13 December 1938; 20 December 1938; 21 July 1939; 14 December 1939; 15 December 1939.

102. For "farm bloc" opposing Boulder Dam and All-American Canal: ibid., 19 October 1922.

103. Mexican Labor in U.S. Field Notes, folder 4, p. 75, carton 10, Series A, Set I, MSS 84/38C, PT.

104. McWilliams, *California,* 101.

105. Kennedy, *Freedom from Fear,* 202. New Deal in the Imperial Valley: Saund, *Congressman from India,* 58–61.

Chapter 4

1. *BN,* 14 July 1920.

2. Farm labor market: Fisher, *Harvest Labor Market.* Roots of California racialized migrant labor: Street, *Beasts of the Field;* Leonard, *Making Ethnic Choices,* chapter 3. Agribusiness control over workers: McWilliams, *Factories in the Fields;* McWilliams, *California;* Guerin-Gonzales, "The International Migration."

3. Mitchell, *Lie of the Land.*

4. Tout, *First Thirty Years,* 78–79, 104–5, 166, 178; *BN,* 4 November 1914; Cory and Blake, *Imperial Valley,* 1331–32; James, *Wonders of the Colorado Desert,* 49–50, 64–65, 503–14. Transition from Indian to white cotton pickers: Mexican Labor in U.S. Field Notes, folder 10, carton 10, Series A, Set I, MSS 84/38C, PT.

5. Majka and Majka, *Farm Workers,* 41–42. Oxnard strike: Almaguer, *Racial Fault Lines,* chapter 7.

6. Taylor, *Mexican Labor,* 5–16. Overview of white itinerant workers: Wyman, *Hoboes.* Wages: *Thirteenth Census, Abstract of the Census, Supplement for California,* 656; Fuller, "Supply of Agricultural Labor," 158, 234–37.

7. Taylor, *Mexican Labor,* 16–17; Mexican Labor in U.S. Field Notes, folder 10, carton 10, PT. Fifty-one laborers imported from China and frequency of strikes: Duncan, "Chinese and Economic Development," 623, 625–26. A white Mexicali grower said Chinese frequently went on strike, which helps explain why farmers switched to campesinos: *BN,* 17 September 1924. Chinese in Mexicali: Hu-DeHart, "Chinese of Baja California"; Gerardo, *El dragón;* Velázquez Morales, *Los inmigrantes chinos.* For 4,000 Chinese on strike in 1918 demanding $4 a day: (ten years ago today) *CC,* 26 June 1928; 5 July 1928.

8. *BN,* 5 October 1914; 10 November 1915. Quote: ibid., 31 August 1918.

9. Quotes: Hall, "Just Like Dixie Land," 175. Imperial Valley black cotton pickers: Duke, "California the Golden," 245, 252, 254.

10. Taylor, *Mexican Labor,* 10–11, 15–16. Racial epithets: *BN,* 22 June 1914. Racial enmity between Mexicans and African Americans: [Nora E. Kreps], *Diary of a Schoolteacher in the Imperial Valley, April 16, 1924,* pp. 3–4, folder 57, "A Survey of Race Relations on the Pacific Coast," 1925, box 24,

SRR. Blacks in urban jobs: *BN,* 10 February 1921; 25 July 1927; 29 September 1933; Kathryn Taylor interview.

11. Howe and Hall, *Story of the First Decade,* 207; Quote: *BN,* 24 May 1912. Description of Japanese working: Heffernan, *Personal Recollections,* 13.

12. Quote: *BN,* 17 May 1920. Assault: ibid., 21 June 1915.

13. Baldoz, *Third Asiatic Invasion;* España-Maram, *Creating Masculinity.* Filipino cotton pickers: *BN,* 26 December 1917.

14. J. L. Leonard, Will J. French, and Simon J. Lubin, United States Special Commission on Agriculture Disturbances in Imperial Valley, *Report to the National Labor Board, February 11, 1934,* p. 20048 (hereafter Leonard Report), *Violations of Free Speech.* Filipinos in Imperial Valley: Delia Gongora interview, 24 November 1999; "A Survey of Race Relations on the Pacific Coast," 1925, pp. 1–2, folder 93, box 25, SRR.

15. Hart, "Mexican Revolution," chapter 14. Revolution's impact on borderlands: Hall and Coerver, *Revolution on the Border,* chapter 8. Welcoming displaced Mexicans: *BN,* 17 September 1918. Labor and transnational Mexican migration: Barton, "At the Edge of Storm"; Barton, "Borderland Discontents"; Cardoso, *Mexican Emigration.*

16. Cardoso, "Labor Emigration."

17. Alvarez, *Familia,* chapters 3–5. Mexican migration and Cantú's road: Aguirre Bernal, *Compendio histórico-biográfico de Mexicali,* chapter 4. Immigration Service: Ettinger, *Imaginary Lines,* chapters 3–5.

18. Ad placed by labor contractor Ayala & Son: "We furnish labor for cantaloupes, lettuce, and general ranch work," *BN,* 5 March 1920. Recruitment of Mexican migratory labor throughout the Southwest: Foley, *White Scourge,* chapter 5. Labor contractors: Peck, *Reinventing Free Labor.* Railroad fares: *BN,* 21 September 1916. Extended discussion of transnational labor: Andrews, *Killing for Coal;* Walsh, *Building the Borderlands.*

19. *Mexicans in California,* chapters 2–3; Guerin-Gonzales, *Mexican Workers;* Reisler, "Always the Laborer"; Taylor, *Mexican Labor,* 7–17; Weber, *Dark Sweat,* 53–62.

20. Reisler, *By the Sweat of Their Brow,* chapters 1–2.

21. *BN,* 3 August 1918; 7 August 1918; 5 December 1918; 7 August 1918; 7 December 1918. Meaning of "white man's country": Jameson, *All That Glitters,* chapter 6, especially 141–43. "White man's camp": Benton-Cohen, *Borderline Americans,* chapter 3.

22. *BN,* 24 September 1918; 6 February 1919. Labor ads: ibid., 12 May 1919.

23. *CC,* 8–12 September 1919; 16–18 September 1919; 15 October 1919; Duncan, "Chinese and Economic Development," 628. Cantú's ban on Chinese: *NYT,* 15 September 1919.

24. O. F. Brant to Ismael Pizarro Su[á]rez, October 28, 1919, telegram, folder Letters 1919, box 291, SL; Ismael Pizarro Su[á]rez to O. F. Brant, October 30, 1919, telegram, folder Letters 1919, box 291, SL. CRLC's lawyer in Mexico City spoke with the minister of the interior, who stated the orders to admit Chinese arrived in Mexicali. The CRLC lawyer also communicated with the Chinese minister through the British legation: Ismael Pizarro Suárez to O. F. Brant, 1 November 1919, folder Letters 1919, box 291, SL. CRLC's follow-up: O. F. Brant to Ismael Pizarro Suárez, 3 November 1919, telegram, folder Letters 1919, box 291, SL. China did not support or facilitate immigration of its citizens to Mexico because of violence against them in Sonora and Sinaloa. The Chinese government "compelled Governor Cantú to keep the word he had already given authorizing the entrance of the 350 men": Ismael Pizarro Su[á]rez to O. F. Brant, 4 November 1919, folder Letters 1919, box 291, SL. The Mexican Department of the Interior sent a telegram to Cantú: "Please admit the 330 [C]hinese, detained in San Francisco, enter national territory; provided they are not comprised within restrictions of migration or sanitary ones": Ismael Pizarro Suárez to O. F. Brant, 6 November 1919, folder Letters 1919, box 291, SL. For all the Chinese crossing the border: O. F. Brant to I. Pizarro Su[á]rez, 29 November 1919, folder Letters 1919, box 291, SL. Violation of treaty with China: Kim, "Immigrants, Workers, Pioneers," 128–35.

25. National Chamber of Agriculture, C[á]mara Agr[í]cola Nacional del Distrito Norte de la Baja Cal. Mexico to Custom Brokers, Calexico, Mexicali, B.C., 1919 [Spanish and English translation], folder Letters 1920, box 291, SL. For Cantú and agricultural chamber: *CC,* 8–10 May 1919. Arnulfo Liera: Gómez Estrada, *Gobierno y casinos,* 53. Arnulfo Liera and Navigation Co.: *CC,* 20 April 1923; 23 January 1924.

26. CRLC evicting campesinos: Herrera Carrillo, *Reconquista y colonización,* 179–180. US employers

requesting Governor Cantú's assistance in procuring Mexican labor: *CC,* 8–9 May 1919; 12 May 1919; 30 July 1919. Circular transmigration: Andrés, "Invisible Borders."

27. *BN,* 2 April 1920. Cotton association's labor policies: ibid., 29 June 1920.

28. Ibid., 20 April 1920. Quote: ibid., 10 April 1920. See also ibid., 14 April 1920; 26–27 April 1920; 3 May 1920.

29. Quote: ibid., 28 April 1920. Ibid., 10 May 1920; 3 May 1920.

30. Ibid., 13 May 1920. Farmers turned away white melon pickers who responded to labor ads in Los Angeles newspapers because employers only wanted "skilled" workers. "Skilled" was code for a Mexican laborer. Construction camps: ibid., 14 May 1920. Cotton association drew up by-laws and negotiated with ginners: ibid., 17 May 1920; 19 May 1920. All seven cotton gins in the Imperial Valley agreed to participate: ibid., 29 May 1920; 3 July 1920. For happiness with program: ibid., 14 July 1920. Deporting seventy-seven "undesirables": ibid., 30 September 1920.

31. Up to that point, the cotton association paid $30,000 to transport Mexicans. The organization had $14,000 in assets, but owed $15,000. A mere $4,600 was collected from farmers with bale tax; banks loaned the association $21,000: ibid., 19 October 1920; 1 November 1920; 15 November 1920. O'Neil: ibid., 11 November 1920.

32. Treaty: Kim, "Immigrants, Workers, Pioneers," 137–38; Duncan, "Chinese and Economic Development," 633. 1923 advertisement: *BN,* 12 October 1923. 13,000 Mexicans: Taylor, *Mexican Labor,* 17–18. 24,000 migrants: de Vivanco, *Baja California,* 327, 415–17, 359. $400,000: *CC,* 6 November 1925.

33. *BN,* 5 May 1920. Shift from higher-paid white workers to inexpensive Mexicans in nonfield jobs: ibid., 1 July 1920; Taylor, *Mexican Labor,* 34.

34. Crop timeline, Leonard Report, 20044–45, *Violations of Free Speech; BN,* 22 October 1932. Peas: CERA Field Notes, Matt & Rugg Field Notes, 1935, folder 40; carton 14, MSS 84/38C, PT. 7,500 harvesters: Taylor, *Mexican Labor,* 35–36.

35. Report: Duke, "California the Golden," 245; Fawcett: folder 72, 1925, pp. 1–3, box 25, SRR. Last quote: Taylor, *Mexican Labor,* 40.

36. For "peon": González, *Culture of Empire.* Quote disparaging Mexicans: Chloe Holt, 27 February 1925, p. 3, folder 297, box 31, SRR.

37. Reisler, *By the Sweat of Their Brow,* 129–30.

38. "Mexican Immigration," no. 50, 5 June 1918, box 4, SRR.

39. Mexican Labor in U.S. Field Notes, p. 75, folder 4, carton 10, Series A, Set I, MSS 84/38C, PT.

40. Taylor, *Mexican Labor,* 40.

41. Mexican Labor in U.S. Field Notes, p. 73, folder 4, carton 10, Series A, Set I, MSS 84/38C, PT.

42. McWilliams, *Factories in the Fields,* 118, 134. Estrada: Guerin-Gonzales, *Mexican Workers,* 56.

43. Animosity between the Japanese and the Mexicans: Andrés, "I Am Almost More at Home," 87. Asian and Mexican relations: Leonard, *Making Ethnic Choices,* 97–98, 137, 140–41.

44. *BN,* 14 October 1918. Truant officers: ibid., 27 January 1920.

45. Duke, "California the Golden," 237–41. Protests against further Mexican importation: *BN,* 24 January 1921.

46. Kakoo Singh and Atma Singh: *BN,* 27 January 1920. Quotes: Duke, "California the Golden," 242. Examples of employing children: Duke, "California the Golden," 238–39; *BN,* 24 January 1921; 3 February 1921.

47. *BN,* 26 October 1925; 1 October 1927. School attendance records: *Fifteenth Census of United States,* 1930 vol. 3, pt. 1, *Population,* 251; *Sixteenth Census of the United States,* 1940 vol. 2, pt. 1, *Population,* 540.

48. 1920s racism, immigration policy, and deportations: Ngai, *Impossible Subjects,* chapters 1–2. 1929 law criminalizing illegal entry: ibid., 60. 1924 immigration policy: Reisler, *By the Sweat of Their Brow,* 12, 59. Border Patrol before World War II: Hernández, *Migra!,* chapters 1–3; Ettinger, *Imaginary Lines,* chapter 6. Mexican Migration Service: Hernández, *Migra!,* chapter 4.

49. *BN,* 29 July 1924. Protest against excessive Mexican laborers: ibid., 1 August 1924.

50. Cramp, Shields, and Thomson, *Study of the Mexican Population,* 4, BL. Testimony from C. B. Moore, US Congress, Senate, *Restriction of Western Hemisphere Immigration,* 62. Complaints about aggressive border patrolling: *BN,* October 28, 1925; November 3, 1925. Lobbying for open border for Mexican laborers: ibid., February 3, 1926; February 6, 1926.

51. Reisler, *By the Sweat of Their Brow,* 60–65. Border Patrol correspondence discussing passbook plan: frames 0766–0775, 0782–0786, 0793–0796, 0828–0837, 0857–0861, reel 15, RG 85.

52. *BN,* 7 April 1926; Taylor, *Mexican Labor,* footnote 47, p. 21. Investigation: Bloch, *Report on the Mexican Labor Situation.* Passbook plan: Cramp, Shields, and Thomson, *Study of the Mexican Population,* 4–5, 118–22.

53. *BN,* 6 July 1920.

54. Ibid., 25 May 1920.

55. Ibid., 22 May 1922. Quote: Taylor, *Mexican Labor,* 41.

56. CERA Field Notes, June 15–July 12, 1935, folder 39, carton 14, MSS 84/38C, PT.

57. Imperial Valley—May 1934, p. 5, folder 30, carton 14, MSS 84/38C, PT.

58. Taylor, *Mexican Labor,* 13–14; Duke, "California the Golden," 250.

59. *BN,* 30 June 1926. Mexican field hand: CERA Field Notes, Marysville to Imperial Valley [March–April 1935], folder 37, carton 14, MSS 84/38C, PT.

60. McWilliams, *Ill Fares the Land,* 19, 44.

61. Employment office: *BN,* 7 August 1926. 193 claims: ibid., 22 November 1926. 1928 law: ibid., 5 April 1928.

62. *Mexicans in California,* 130–33. Example of complaints: *BN,* 28 July 1920.

63. Louis Bloch, *Report on the Strike of the Imperial Valley Cantaloupe Pickers, 26 May 1928,* folder 11, carton 46, MSS 84/38C, PT (hereafter *Report on the Strike*); *Mexicans in California,* 131, 133.

64. *Mexicans in California,* 131, 133; *BN,* 17 July 1929.

65. *Report on the Strike,* 2–3. The state could not compel intransigent employers to pay and contractors disappeared, ibid., 6.

66. Taylor, *Mexican Labor,* 55–58, 65–69; Duke, "California the Golden," 233–37. Example of workers forced to camp alongside fields because of a lack of housing: *BN,* 19 June 1920. Farm leader admitting horrible farm labor housing: ibid., 10 April 1926.

67. Holliday, "A Study of Non-Attendance," 8.

68. Westside: Kathryn Taylor interview; Modesto Ibarra interview. Filthy animals: Imperial Valley—May 1934, p. 5, folder 30, carton 14, MSS 84/38C, PT.

69. Mexican Labor in U.S. Field Notes, October 1928, p. 77, folder 4, carton 10, Series A, Set I, MSS 84/38C, PT. Mexican laborer said meager wages compelled his compatriots to build multiple abodes on urban property and this degraded sanitation: CERA Field Notes, June 15–July 12, 1935, folder 39, carton 14, MSS 84/38C, PT.

70. *BN,* 10 April 1926.

71. Ibid., 6 June 1930. Description of town housing: Elliott, "Health and Relief Problems," 32–33.

72. Mexican Labor in U.S. Field Notes, October 1928, p. 77, folder 4, carton 10, Series A, Set I, MSS 84/38C, PT. Quote: Bloch, *Report on the Mexican Labor Situation,* 116.

73. Quote: *BN,* 11 July 1918. Commission of Immigration and Housing struggle against agribusiness: Daniel, *Bitter Harvest,* 88–101.

74. *BN,* 16 June 1926. Racist ideology of California Commission of Immigration and Housing and Rugg's showdown with agribusiness: Mitchell, *Lie of the Land,* 102–7. State camp inspectors bringing suit against white, Mexican, and South Asians: *BN,* 8 January 1929. Worst camps in state: ibid., 2 July 1931. Quote: Leonard Report, 20047, *Violations of Free Speech.* Labor camp inspections: *BN,* 18 April 1934.

75. Traumatized federal labor investigator: Campbell MacCulloch [Los Angeles Regional Labor Board] to Mrs. B. M. Stern, National Labor Board, Washington, D.C., February 24, 1934, folder 6, General Correspondence-NLRB Offices, Nov. 1933–June 1934, box 2, RG 25. "Gaunt": McWilliams, *Factories in the Fields,* 131–32. Racism and violence against Filipinos: Andrés, "Power and Control," 362–63, 382, 385–86. Overview: Baldoz, *Third Asiatic Invasion,* chapters 3–4.

76. *BN,* 13 June 1934.

77. Stein, *California and the Dust Bowl,* 45; Gregory, *American Exodus,* chapters 1–2. "Okies": Andrés, "Power and Control," 241–42.

78. Weber, *Dark Sweat,* chapter 5; McWilliams, *Factories in the Fields,* 322.

79. McWilliams, *Factories in the Fields,* 189; Leonard Report, 20046, *Violations of Free Speech; BN,* 1 November 1933. Quote: Guerin-Gonzales, "International Migration," 167. Pérez: *BN,* 22 June 1939.

80. Repatriation from the US perspective: Balderrama, *Decade of Betrayal.* Repatriation from Mexico's perspective: Alanís Enciso, *Que se queden allá.* Mexican functionary criticized foreign ownership of the Mexicali Valley: Spanish and English translation of excerpt, *Report Submitted by General Juan*

Andrew Almaz[á]n, Secretary of Communications and Public Works, to the President of the Republic, with Reference to His Journey through the Northwestern Part of the Country, Mexico, July, 1930, 29–37, folder Mexico, file 1, IID. Land expropriation: Dwyer, *Agrarian Dispute.* American investors in Mexico's cotton industry: Walsh, *Building the Borderlands.* 1930 quota hearings: Testimony from Harry Chandler, US Congress, House, *Western Hemisphere Immigration,* 59–74; testimony from Chester Moore, ibid., 239–251. Quotes: *BN,* 1 February 1930.

81. Taylor, *Mexican Labor,* 6. South Asians thrown into field labor: Leonard, *Making Ethnic Choices,* 139–41.

82. *BN,* 4 November 1929. Pressure to employ whites: ibid., 23–24 September 1929; 27–28 September 1929. Pressure to hire white migrants: 1–3 October 1929; 31 October 1929. White spot: ibid., 24 December 1929; 29 March 1930; 28 July 1930; 27 January 1932.

83. Ibid., 27 February 1930. Hoboes and crime: ibid., 25 April 1930; 28 April 1930; 26 December 1930; 2 January 1931.

84. Leonard Report, 20046–20048, *Violations of Free Speech.* Migratory life during Depression: Walton, "Study of Migratory Mexican Pea-Pickers," 44–48, 61–62, 65–66, 93–102, 105, 127, 132, 141–43, 161, 183.

85. CERA Field Notes, June 15–July 12, 1935, folder 39, carton 14, MSS 84/38C, PT.

86. CERA Field Notes, Marysville to Imperial Valley [March–April 1935], folder 37, carton 14, MSS 84/38C, PT.

87. Gregory, *American Exodus,* chapter 5. Quote: CERA Field Notes: Marysville to Imperial Valley [March–April 1935], folder 37, carton 14, MSS 84/38C, PT.

88. McWilliams, *California,* 154, 159.

89. *BN,* 26 December 1930. Later reference to private relief: ibid., 17 November 1931.

90. Ibid., 1–2 December 1931; 28 December 1931. 250 labor camps: McWilliams, *Factories in the Fields,* 296. 1932–33 state labor camp statistics: *BN,* 4 October 1933. Mexicali food riot: two articles *NYT,* 22 February 1931. Large-scale family protests in Mexico during the Great Depression: Olcott, *Revolutionary Women in Postrevolutionary Mexico,* chapter 4.

91. Mutualistas: *BN,* 20 April 1925. Bloch's quote: ibid., 10 April 1926. Mexicans in Calexico paying for the health care of Mexican expatriates, testimony from C. B. Moore, US Congress, House, *Immigration from Countries of the Western Hemisphere,* 286; testimony from C. B. Moore, US Congress, Senate, *Restriction of Western Hemisphere Immigration,* 61. For heartbreaking effort by La Cruz Azul Mexicana in Brawley to raise money: *BN,* 13 October 1932; 15 October 1932; 20–21 October 1932; 16 November 1932; 18 November 1932. Japanese paid medical care of destitute countrymen: ibid., 21 June 1938. Ideology of Japanese self-help: Takanori (Pro) Nimura interview, 19 March 1998. Public charge: Molina, "Constructing Mexicans"; Aguila, "Protecting 'Mexico de Afuera,'" 65–82.

92. *BN,* 27 January 1932; 7 September 1932; 12 September 1932; 26 September 1932; 29 September 1932; 5 October 1932; 19 September 1932; 10 November 1932. 1.2 million: Starr, *Endangered Dreams,* 227. 1932 aid statistics: *BN,* 16 January 1933; 21 January 1933.

93. *BN,* 8 December 1932; 18–19 January 1933; 22 April 1933; 8 June 1933; 16 January 1933. Editorial: ibid., 8 March 1933. Collapse of self-help: ibid., 11–12 July 1933.

94. Stein, *California and the Dust Bowl,* 82. Depression-era California relief and New Deal projects: Starr, *Endangered Dreams,* chapters 8, 10–12; Stein, *California and the Dust Bowl,* chapters 5–6; McWilliams, *Factories in the Fields,* 285–304. RFC and WPA depended on local charity committees and county officials for project and policy recommendations: *BN,* 1 November 1933. Residency requirements: ibid., 11 July 1933.

95. *BN,* 13–14 November 1933. Struggle to maintain local control of relief to benefit farming: ibid., 1 November 1933; 9 November 1933; Imperial Valley—May 1934, p. 1, folder 30, carton 14, MSS 84/38C, PT.

96. *BN,* 24–25 October 1933; 20 November 1933.

97. San Joaquin strike: Weber, *Dark Sweat,* chapter 3. Denial of aid who refused a field job: *BN,* 13–14 November 1933. Federal money: ibid., 25 October 1933; 25 November 1933; 27 November 1933; 2 December 1933; 4 December 1933; 13 December 1933; 7 December 1933; 19 December 1933. Nonresident work projects: ibid., 20 December 1933; 30 December 1933; 28 December 1933.

98. Job cutbacks: ibid., 20 January 1934. Editorial: ibid., 16 February 1934.

99. Calipatria: ibid., 14 February 1934; 28 February 1934. 9,895 receiving groceries: ibid., 2 May 1934. All-American Canal jobs in October: ibid., 25 October 1934; 31 October 1934; 3 January 1935; 27 February 1935.

100. Mitchell, *Campaign of the Century;* Starr, *Endangered Dreams,* chapter 5; McWilliams, *California,* 161; McWilliams, *Ill Fares the Land,* 7.

101. McWilliams, *Factories in the Field,* chapter 1; McWilliams, *California,* 162; McWilliams, *Ill Fares the Land,* chapter 1. For agribusiness counterattack against federal assistance to strikers: *BN,* 3 January 1935; McWilliams, *Factories in the Fields,* 287–96. FTS: Starr, *Endangered Dreams,* chapter 8.

102. SERA: Starr, *Endangered Dreams,* chapter 8. 600 families: McWilliams, *Factories in the Fields,* 289–93. Criticism of SERA: *BN,* 13 March 1935; 15–16 March 1935; 18 March 1935; 22 March 1935. Best of both worlds: ibid., 1 April 1935; 24 April 1935; 1 May 1935.

103. Ibid., 8–9 July 1935. SERA's definition of unemployable and row over aid: ibid., 10–11 July 1935; 22 July 1935. Second transfer: ibid., 15 October 1935; 23 October 1935; 13 November 1935; 2 December 1935.

104. End of federal relief: ibid., 6 January 1936; 19 February 1936; 26 June 1936. Mexicans dropped from relief rolls during harvest: ibid., 29 March 1937. Charity or indigent cases: ibid., 13 November 1937. Lowest indigent expense: ibid., 2 March 1937; 13 November 1937; 30 November 1937; 17 June 1938. Agribusiness was satisfied with relief subsidy: Stein, *California and the Dust Bowl,* 82–84, chapter 5. Board of Supervisors denying Mexican immigrants access to health care in defiance of state law: Elliott, "Health and Relief Problems," 53.

105. *BN,* 25 July 1935; H. E. Drobish to Dr. Paul Taylor, 5 October 1935, folder 25, carton 14, MSS 84/38C, PT; Starr, *Endangered Dreams,* chapter 9.

106. Stein, *California and the Dust Bowl,* 110, 116, 129. Letter: *BN,* 20 January 1939. Under Olson, SERA's relief load rose 40 percent higher than previous year: ibid., 22 August 1939. Still it was difficult to get relief. The Imperial Valley SERA office rejected half its applicants: "A man or woman," asserted its director, "has to prove absolute destitution before acceptance for state relief," ibid., 30 October 1939.

107. Conservative backlash: McWilliams, *Ill Fares the Land,* 43–45. Quote: Majka and Majka, *Farm Workers,* 119; Stein, *California and the Dust Bowl,* 117, chapter 4; Burke, *Olson's New Deal.*

108. *BN,* 27 November 1939; 29 December 1939.

Chapter 5

1. Shed strikes: lettuce, *BN,* 29 November 1920; 22 December 1920; cantaloupe, ibid., 7–8 June 1921; lettuce, ibid., 5 December 1921; 12 December 1921.

2. Rural working-class radicalism: Preston, *Aliens and Dissenters;* Weber, *Dark Sweat;* chapter 2.

3. Daniel, *Bitter Harvest;* Majka and Majka, *Farm Workers.*

4. Hall, *Harvest Wobblies.* Southwest: Mellinger, "How IWW Lost." California: Starr, *Endangered Dreams,* chapter 2.

5. Blaisdell, *Desert Revolution,* 39. Wobblies in Imperial Valley: *BN,* 16 April 1912.

6. Starr, *Endangered Dreams,* 34–44, 48; Whitten, *Criminal Syndicalism,* 6–7; Majka and Majka, *Farm Workers,* 53–57; Daniel, *Bitter Harvest,* 96–99.

7. Calexico: *BN,* 24 February 1914; 26 February 1914. Imperial: ibid., 26 January 1914. Brawley: ibid., 27 July 1914. Brawley: ibid., 17 March 1915. El Centro: ibid., 23 April 1915.

8. Wobbly agricultural organizing: Hall, *Harvest Wobblies,* chapter 3. Wobblies in Imperial Valley: ibid., 69, 95; Daniel, *Bitter Harvest,* 84–85. Wobblies in Brawley: *BN,* 20 February 1915; 24 February 1915; 6 June 1917. Ten Wobblies: ibid., 23 January 1917. *BN* followed IWW activity nationwide: ibid., 8 June 1914; 17 February 1915; 11 September 1915; 11–12 July 1917; 5 November 1917; 26 December 1917; 22 April 1920.

9. Majka and Majka, *Farm Workers,* 58; Starr, *Endangered Dreams,* 48–50. Syndicalism law outlawed "advocating, teaching or aiding and abetting the commission of crime, sabotage . . . or unlawful acts of force and violence or unlawful methods of terrorism as a means of accomplishing a change in industrial ownership or control, or affecting any political change," Whitten, *Criminal Syndicalism,* 65.

10. El Centro: *BN,* 20 April 1918. Calipatria: ibid., 9 September 1919; 11 September 1919. Blake: ibid., 13 March 1920; 26 March 1920.

11. Ibid., 31 May 1921; 23 June 1921; 29 July 1921.

12. *BN* did not cover trial or appeal: ibid., 6 September 1923. Additional Wobbly references: R. Leonard charged with criminal syndicalism: ibid., 19 December 1921. A few months later, a "WHOLE NEST OF I.W.W.'S FOUND BY AGENT HOWES": ibid., 1 February 1922; also ibid., 7 February 1922; 23 February 1922.

13. Ibid., 5 June 1919.

14. Revolutionary violence on US-Mexico borderlands: Sandos, *Rebellion in the Borderlands;* Hall and Coerver, *Revolution on the Border.* Bracero militancy in northern Mexico: Santiago, *Ecology of Oil;* Calderón, *Mexican Coal Mining.*

15. H. H. Clark to Harry Chandler, February 20, 1921, folder Letters 1921, box 291, SL.

16. CRLC's struggle with squatters: Kerig, "Yankee Enclave," 118–20.

17. Camp John H. Beacom, Calexico: *CC,* 14 May 1919; 21 June 1919. US annexation efforts in Mexico: Stout, *Schemers and Dreamers.* Quote: *BN,* 9 August 1920. November 1919 military attack plan: Kirchner, *Baja California Railways,* 72–79. Execution by firing squad: *NYT,* 19 November 1921.

18. Sánchez Ramírez, *Crónica agrícola,* 82–83.

19. Mexican Labor in U.S. Field Notes, p. 68, folder 4, carton 10, Series A, Set I, MSS 84/38C, PT.

20. *Report Rendered to the Honorable Federal Board of Investigation by La Comit[é] Mexicano de Bienestar Social (Mexican Social Welfare Committee) of Brawley, California, February 1934,* folder 38, carton 14, MSS 84/38C, PT (hereafter *Report Rendered to the Honorable Federal Board*).

21. *CC,* 25 May 1922; 29 May 1922.

22. *BN,* 24 May 1922; 1 June 1922 (from *CC*). Japanese: ibid., 24 May 1922; 26 May 1922.

23. Quotes: ibid., 26 May 1922 (from *CC*). Martial law and detaining agitators: *CC,* 23–24 June 1922; 1 December 1922. Mexican leaders in 1920s using troops to smash *agraristas:* Tobler, "Peasants and Shaping."

24. *Report Rendered to the Honorable Federal Board,* n.p.

25. General Abelardo Rodríguez evicted squatters: Sánchez Ramírez, *Crónica agrícola,* 83–84. Governor Abelardo Rodríguez transported unemployed braceros out of the Mexicali Valley: *CC,* 1 April 1925.

26. *BN,* 29 November 1922; 31 January 1924; 30 October 1925.

27. For the Mexican government using its consuls to control Mexican expatriates: González, *Mexican Consuls,* chapter 2; González, "Company Unions"; Sánchez, *Becoming Mexican American,* chapter 5.

28. In addition to previously listed sources: Snodgrass, '"We Are All Mexicans Here"; Hart, "Mexican Revolution"; Middlebrook, *Paradox of Revolution,* chapter 3.

29. Suarez-Potts, *Making of Law.* Example of union wage scale with Colorado River Land Company: *CC,* 10 June 1927. Example of Dr. Ernesto Ostos, commissioner of labor for the Northern District of Baja California: ibid., 31 October 1928.

30. *Mexicans in California,* 129. CUOM: ibid., chapter 6. Critical analysis of CUOM: González, *Mexican Consuls,* 65–81.

31. Mutualistas in this union's leadership: Hernández, *Mutual Aid for Survival,* 76–79; Louis Bloch, *Report on the Strike of the Imperial Valley Cantaloupe Pickers, 26 May 1928,* folder 11, 9–11, carton 46, MSS 84/38C, PT (hereafter *Report on the Strike*).

32. Circular 18 May 1928, folder 19, carton 10, MSS 84/38C, PT; González gave Paul Taylor union documents, letter 27 December 1928, folder 19, carton 10, MSS 84/38C, PT.

33. Circular 16 April 1928, folder 19, carton 10, MSS 84/38C, PT; *BN,* 16 May 1928. Carlos Ariza: González, *Mexican Consuls,* 23–24, 32, 166. Law firm: *Report on the Strike,* 5–6.

34. Circular 18 May 1928, folder 19, carton 10, MSS 84/38C, PT. Bloch updated his report in December to include comments from District Attorney Heald and summary of reforms in labor contract system: *Mexicans in California,* 149–50. Letter: Filemon B. González to Hermolao E. Torres, 18 April 1928, upper case in original; Spanish and English translation, folder 11, Exhibit 2, carton 46, MSS 84/38C, PT.

35. Circular 25 April 1928, folder 19, carton 10, MSS 84/38C, PT; articles of incorporation: letters 8 May 1928 and 13 May 1928, folder 19, carton 10, MSS 84/38C, PT. Discussion of articles of incorporation: *BN,* 9 May 1928; *Report on the Strike,* 7–8. Society members wanted to call the union the "Mexican Workers Union of the Imperial Valley," but Ariza suggested following strategies from the Japanese

growers association, like selecting a nonethnic name, circular 18 May 1928, folder 19, carton 10, MSS 84/38C, PT; laborers paid $1 to join and $1 every month afterward.

36. Letter 13 May 1928, folder 19, carton 10, MSS 84/38C, PT; Mexican Labor in U.S. Field Notes, 18 October 1928, p. 71, folder 4, carton 10, Series A, Set I, MSS 84/38C, PT; demands, folder 19, Exhibit 4, carton 10, MSS 84/38C, PT. Some union demands and farmers' reaction: *Report on the Strike,* 7–11; *BN,* 26 May 1928.

37. Letter 5 May 1928, p. 12, folder 19, carton 10, MSS 84/38C, PT.

38. Ibid., 13–15; *BN,* 6 June 1928.

39. *BN,* 8 May 1928. One flyer, "in illiterate Spanish," informed braceros that shippers had sent a representative to Washington, DC, to fight the Box bill, which sought to put an immigration quota on Mexico. The union supported the quota because unlimited migration permitted growers to replace strikers with new workers, ibid. Chairman: Wollenberg, "Huelga," 54. Operatives: *BN,* 8 May 1928. Flyers from farmers: *Report on the Strike,* Exhibits 7–9. Translations of flyers: *Report on the Strike,* 25–27. Flyers: *BN,* 26 May 1928.

40. Letter 8 May 1928, folder 19, carton 10, MSS 84/38C, PT.

41. Quotes in order: *BN,* 9–10 May 1928; 9 May 1928.

42. Arresting union leaders: *Report on the Strike,* 15. Judges released workers: ibid., 15–16; *BN,* 17 May 1928. Noose: ibid., 10 May 1928.

43. *BN,* 11 May 1928.

44. Ibid., 14 May 1928; 15 May 1928; 16 May 1928.

45. Letter: 13 May 1928, 1, folder 19, carton 10, MSS 84/38C, PT.

46. 16 May 1928, folder 19, carton 10, MSS 84/38C, PT; *El Nuevo Mundo,* folder 11, Exhibit 6, carton 46, MSS 84/38C, PT; translated news article, Viernes 11 de Mayo de 1928, folder 19, carton 10, MSS 84/38C, PT.

47. Releasing prisoners: *BN,* 12 May 1928; 15 May 1928. Interpreter: *Report on the Strike,* 18–20; *BN,* 16 May 1928. Mexican immigrants feared guilty plea would be used to deport them: *BN,* 16 May 1928; *Report on the Strike,* 18–20, 27, endnote 17, 28. Government protest: González, *Mexican Consuls,* 167–68.

48. *Report on the Strike,* 29. Quotes: ibid., 8, 29–31.

49. Taylor's Imperial Valley 1927 field notes, folder 10, carton 10, MSS 84/38C, PT; handwritten and typed notes, Mexican Labor in U.S. Field Notes, folders 4–5, carton 10, Series A-B, Set I, MSS 84/38C, PT. Quotes in order: Mexican Labor in U.S. Field Notes, 28 October 1928, p. 68, 71, 73–74, folder 4, carton 10, Series A, Set I, MSS 84/38C, PT.

50. Quotes in order: Mexican Labor in U.S. Field Notes, 28 October 1928, pp. 71–77, folder 4, carton 10, Series A, Set I, MSS 84/38C, PT.

51. Bank remittances to Mexico, Bank Statement 31 May 1927, folder 20, carton 10, MSS 84/38C, PT.

52. Packers refused to walk out in solidarity with the Mexican union: *BN,* 12 May 1928. Japanese shippers remained neutral: ibid., 26 May 1928. Bloch investigation: ibid., 4 June 1928; 6 June 1928. New consul: ibid., 7 June 1928.

53. *Report Rendered to the Honorable Federal Board,* n.p. Bloch's labor reform efforts: *BN,* 25 April 1929; 3 May 1929. Licensing fee: ibid., 7 January 1931. State labor commissioner ruling in favor of workers in wage holdbacks: ibid., 6 July 1929. "Negro" labor contractor arrested in Los Angeles for defrauding workers: ibid., 25 July 1927. Lawsuit against labor contractor Archie Ferrera who "absconded" with $1,444: ibid., 9 January 1930. Fred Catalina, Filipino labor contractor, arrested in Los Angeles: ibid., 25 May 1931. Testimony from Chester Moore, US Congress, House, *Immigration from Countries of the Western Hemisphere,* 272–93.

54. *BN,* 27 December 1929.

55. González, *Mexican Consuls,* 169; *BN,* 28 December 1929. Consul's arbitration request: González, *Mexican Consuls,* 51–53; Daniel, *Bitter Harvest,* 111. Chester Moore of Western Growers stated individual farmers did not have to accept agreement: *BN,* 31 December 1929. Red smear: ibid., 2–3 January 1930; 28 January 1930; 4 February 1930. Blaming consul: Peters, "Agricultural Unionism," 41.

56. Daniel, *Bitter Harvest,* 109–12. In 1934, Brawley's Young Communist League had fifteen members: Weber, *Dark Sweat,* 85–88. Harvey's quote: Daniel, *Bitter Harvest,* 113. AWIL's demand to speak: *BN,* 10 January 1930; 13 January 1930.

57. Spector, *Story of the Imperial Valley,* 18. Leaflets: Starr, *Endangered Dreams,* 67–68. Division

among workers: González, *Mexican Consuls,* 171; Spector, *Story of the Imperial Valley,* 19–20. Arrest and charges against communist leaders: *BN,* 14 February 1930; 18 February 1930; 27 February 1930; Daniel, *Bitter Harvest,* 115.

58. González, "Company Unions," 57–58. Mejía: *BN,* 14 January 1930. Strike ends: Spector, *Free the Imperial Valley Prisoners,* 13–14; Spector, *Story of the Imperial Valley,* 19–20; Daniel, *Bitter Harvest,* 116.

59. Voluntary repatriation: Andrés, "Invisible Borders." Ignoring land petitions: ibid., 9–10.

60. *Agrarista* movement, arrest and banishment to Islas Marías: *CC,* 17 October 1927; Garduño, *La disputa,* 85–90; Acosta Montoya, "Precursores del agrarismo," 45–57; Sánchez Ramírez, *Crónica agrícola,* 102–4.

61. Rejection of AWIL: Spector, *Free the Imperial Valley Prisoners,* 13–14; Spector, *Story of the Imperial Valley,* 19–20; Daniel, *Bitter Harvest,* 116. Packer strike: *BN,* 12–13 February 1930; Spector, *Story of the Imperial Valley,* 21–22. Filipinos and conservative Mexican union: *BN,* 14–15 February 1930; 17 February 1930. Conciliator: ibid., 19 February 1930; Daniel, *Bitter Harvest,* 117–118. Strike's end: *BN,* 28 February 1930.

62. *BN,* 20 February 1930; 28 February 1930; 5 March 1930; 7–8 March 1930; 25 March 1930; 1 April 1930. In-depth discussion: Gray, *American Civil Liberties.*

63. Spector, *Free the Imperial Valley Prisoners,* 14–15, 20. Definition of disturbing the peace: *BN,* 22 September 1931. Repression: Spector, *Story of the Imperial Valley,* 5–6, 23–24; Spector, *Free the Imperial Valley Prisoners,* 15. Spies: Daniel, *Bitter Harvest,* 119–20. District attorney justified arrests: *BN,* 15 April 1930. Thirty-four people arrested for criminal syndicalism: Spector, *Story of the Imperial Valley,* 5–6; Spector, *Free the Imperial Valley Prisoners,* 15; *BN,* 16–17 April 1930; 19 April 1930. Background of defendants: *BN,* 21–24 April 1930; 23 May 1930. Strike collapse and court case: ibid., 31 May 1930; 3 June 1930; 6 June 1930; 9–11 June 1930. Trial: ibid., 13 June 1930; Spector, *Story of the Imperial Valley,* 25–29. District attorney's spies: *BN,* 4–6 June 1930. Trial and appeal: Daniel, *Bitter Harvest,* 123–26; *BN,* 10 June 1930; 13 June 1930. Verdict, sentencing, and ACLU response: ibid., 14 June 1930; 16–17 June 1930; 28 June 1930; 15 September 1930; 27 September 1930; Gray, *American Civil Liberties,* 17; Daniel, *Bitter Harvest,* chapter 4, note 63; Spector, *Free the Imperial Valley Prisoners,* 16.

64. Quote: Spector, *Story of the Imperial Valley,* 28. Consul: González, *Mexican Consuls,* 172–73.

65. McWilliams, *Ill Fares the Land,* 15; Starr, *Endangered Dreams,* chapter 3, p. 71–73; Majka and Majka, *Farm Workers,* 74–87; Daniel, *Bitter Harvest,* 134–35; *BN,* 5 August 1933. San Joaquin strike: Weber, *Dark Sweat,* chapter 3; Daniel, *Bitter Harvest,* chapter 5. A strike sign: "Negro and white unite in fight," *BN,* 10 October 1933.

66. *Report Rendered to the Honorable Federal Board,* n.p.

67. Ibid.; Daniel, *Bitter Harvest,* chapter 7; *BN,* 16 October 1933; 17 October 1933; Leonard Report, 20043–52, *Violations of Free Speech.*

68. Quote: *BN,* 17 October 1933; also ibid., 10 May 1934; 20 June 1934. Terrazas: González, *Mexican Consuls,* chapter 5.

69. Gray, *American Civil Liberties,* chapter 3; *BN,* 21 November 1933.

70. Starr, *Endangered Dreams,* 80–81; Healey, *California Red,* 44. Azteca complex: *Report Rendered to the Honorable Federal Board,* n.p.

71. Leonard Report, 20043–44, *Violations of Free Speech.* Quote: *Report Rendered to Honorable Federal Board,* n.p. Peters, "Agricultural Unionism," 71; Starr, *Endangered Dreams,* 80–81; *BN,* 8–9 January 1934; Walton, "Study of Migratory Mexican," 134, 198.

72. *BN,* 10 January 1934; 12 January 1934. Shed union representative rejected CAWIU appeal to address packers: "Strike," folder 11, n.d., carton 46, MSS 84/38C, PT. Quote: Campbell McCullough, *Labor Conditions in Imperial Valley,* 20039 (hereafter McCullough Report), *Violations of Free Speech.* Board of Supervisors: Gray, *American Civil Liberties,* 29.

73. "amazons": *BN,* 15 January 1934. 115 picketers: ibid., 10 January 1934. Terrazas: González, "Company Unions," 62; *BN,* 10–13 January 1934; 15 January 1934; 18 January 1934; 20 January 1934.

74. *Report Rendered to the Honorable Federal Board,* n.p.; *BN,* 12 January 1934; González, *Mexican Consuls,* 179; McCullough Report, 20038, *Violations of Free Speech;* Gray, *American Civil Liberties,* 228–29; Healey, *California Red,* 48–49. Support for assault on Azteca Hall: *BN,* 13 January 1934; 19 January 1934. Ransacking the hall and civil liberties violations: *Report Rendered to the Honorable Federal Board,* n.p.; *BN,* 12 January 1934; Healey, *California Red,* 49. Legionnaire vigilantes: *BN,* 15 January 1934.

75. *Report Rendered to the Honorable Federal Board,* n.p. Quote: *BN,* 13 January 1934.

76. *BN,* 18 January 1934; 23 January 1934. Names and ages persons arrested: ibid., 13 January 1934. Strike suppression: ibid., 15–17 January 1934; 22 February 1934; 19 April 1935.

77. McCullough Report, 20037, 20040–20041, *Violations of Free Speech.* District attorney was the only significant official who was not a farm landowner: ibid., 20039.

78. Healey, *California Red,* 49; *BN,* 19–20 January 1934; 23 January 1934; 25 January 1934. Habeas corpus: ibid., 16–20 January 1934; 24 January 1934; 28 February 1934. ACLU publicity and telegram to governor: ibid., 18 January 1934. Azteca Hall meeting and speakers: ibid., 18 January 1934; 20 January 1934; 22 January 1934. A. L. Wirin's criminal complaint against police chiefs: ibid., 23 January 1934.

79. *BN,* 24–25 January 1934; Leonard Report, 20044, *Violations of Free Speech;* A. L. Wirin to Ira B. Cross, 13 February 1934, folder 51, carton 14, MSS 84/38C, PT. Sexually assaulted woman counted twenty-seven guns: Gray, *American Civil Liberties,* 31–32; *BN,* 24–25 January 1934; Healey, *California Red,* 48. Wirin's telegram to President Franklin Roosevelt: *BN,* 24 January 1934. Wirin demanded grand jury investigation: ibid., 2 February 1934; also Williams, "Imperial Valley Mob." Wirin filed two lawsuits: *BN,* 24–25 January 1934; 19 July 1934; 9–10 February 1934; Gray, *American Civil Liberties,* 33–35. Attorney Grover Johnson filed civil lawsuit for false imprisonment: *BN,* 23 January 1934.

80. *BN,* 29 January 1934; 31 January 1934; 5 February 1934; Gray, *American Civil Liberties,* 35. Investigators Dr. J. Leonard, Simon Lubin, Will French: *BN,* 27 January 1934. Opinion the NRA did not have jurisdiction: ibid., 25 January 1934; 7 February 1934. Quotes: Leonard Report, 20048, *Violations of Free Speech.*

81. *BN,* 27 February 1934. Communist criticism: Gray, *American Civil Liberties,* 41–42. La Comité: *BN,* 29 January 1934; 1 February 1934. La Comité's report discussed the 1922, 1928, 1930, 1934 strikes and made recommendations to end abuses: *Report Rendered to the Honorable Federal Board,* n.p.

82. *BN,* 9–10 February 1934. "Poor white trash": Imperial Valley—May 1934, p. 1, folder 30, carton 14, MSS 84/38C, PT. Radicalization of "Okies": Stein, *California and the Dust Bowl,* chapters 8–9; Gregory, *American Exodus,* chapter 3. "Red revolution": Gray, *American Civil Liberties,* 43. Liberals wanted the state highway patrol to enforce civil liberties: A. L. Wirin to Ira B. Cross, 13 February 1934, folder 51, carton 14, MSS 84/38C, PT; Ira B. Cross to Raymond V. Cato, 20 February 1934, folder 51, carton 14, MSS 84/38C, PT; E. Raymond Cato to Ira B. Cross, 24 February 1934, folder 51, carton 14, MSS 84/38C, PT. Feeding strikers: *BN,* 12–14 February 1934.

83. "American blood": *BN,* 15 February 1934. Cut off my hand: ibid., 20 February 1934. CAWIU organizers: Healey, *California Red,* 52. County ordinance: *BN,* 16–17 February 1934.

84. Feature article: *BN,* 19 February 1934. Antipicketing ordinance: ibid., 16 April 1934; Gray, *American Civil Liberties,* 43–44.

85. Imperial Valley—May 1934, folder 30, n.p., carton 14, MSS 84/38C, PT.

86. Health board: *BN,* 19 February 1934. Burning camp: ibid., 20 February 1934. Names of picketers: ibid., 19 February 1934; also ibid., 21 February 1934; 10 March 1934; Gray, *American Civil Liberties,* 44. Losing money and convicting CAWIU leaders: *BN,* 27 February 1934; 7 March 1934; 9 March 1934; 12 March 1934; 14–15 March 1934; Healey, *California Red,* 53–54.

87. *BN,* 20 February 1934; 23 February 1934; 6 March 1934; 8–9 March 1934. Communication between National Labor Relations Board, Terrazas, and Alvin Jack: González, "Company Unions," 66. Luis Valdés: *BN,* 23 February 1934.

88. Williams, "Imperial Valley Prepares for War"; González, *Mexican Consuls,* 187. CAWIU leader arrested: Healey, *California Red,* 44. "Placing us at the mercy": González, *Mexican Consuls,* 195. Quote from Mexicali Labor Board representative: *BN,* 12 March 1934. Days later, Terrazas met with handpicked men to perfect his union, which was named La Asociación: ibid., 15 March 1934; also "Boletin," folder 11, n.d., carton 46, MSS 84/38C, PT; "IMPORTANT EVENTS DURING FEBRUARY AND MARCH 1934," folder 11, carton 46, MSS 84/38C, PT; *BN,* 27–28 March 1934.

89. *BN,* 15 March 1934; 21 March 1934; 23 March 1934. Associated Farmers and others: Chambers, *California Farm Organizations,* 8–9, 12. Associated Farmers wielding power: McWilliams, *Ill Fares the Land,* chapter 1; McWilliams, *Factories in the Field,* chapter 1; McWilliams, *California.* Charles Nice, Brawley American Legion commander: *BN,* 27 October 1920. Nice, president of Anti-Communism Association: ibid., 13 March 1934; 4 April 1934. Nice county indigent commissioner, ibid., 8–9 July 1935. Quote: ibid., 22 March 1934; also ibid., 4 January 1935.

90. *BN,* 15 March 1934; Leonard Report, 20051–20052, *Violations of Free Speech; BN,* 21 March 1934; 23 March 1934. AFL field union and quote: Imperial Valley—May 1934, folder 30, carton 14, MSS 84/38C, PT.

91. Williams, "Imperial Valley Prepares for War," 199; Gray, *American Civil Liberties,* 50. Reporting Azteca Hall speeches: *BN,* 19 March 1934. Alvin Jack's quote: "IMPORTANT EVENTS DURING FEBRUARY AND MARCH 1934," folder 11, carton 46, MSS 84/38C, PT. Kidnappings and takeover of Calexico meeting: *BN,* 26–27 March 1934; Jones, "Kidnap Valley." *BN* printed names and car license plate numbers of ACLU lawyers: *BN,* 26–27 March 1934; 17 August 1934; 14 December 1934.

92. "IMPORTANT EVENTS DURING FEBRUARY AND MARCH 1934," folder 11, carton 46, MSS 84/38C, PT; Imperial Valley—May 1934, p. 5, folder 30, carton 14, MSS 84/38C, PT.

93. Gray, *American Civil Liberties,* 55. Opposition to Glassford's appointment: *BN,* 29 March 1934; also "IMPORTANT EVENTS DURING FEBRUARY AND MARCH 1934," folder 11, carton 46, MSS 84/38C, PT. Agribusiness committee: *BN,* 2–3 April 1934.

94. *BN,* 7 April 1934; 9–10 April 1934; 12 April 1934; 16 April 1934. Bailey: Gray, *American Civil Liberties,* 61. Glassford's plan: ibid., chapters 6–9.

95. Quote: Gray, *American Civil Liberties,* 63. Agribusiness report, which Associated Farmers printed and distributed copies of: John Phillips, C. B. Hutchinson, W. C. Jacobsen, The *Imperial Valley Farm Labor Situation, 16 April 1934,* folder 11, 14–16, carton 46, MSS 84/38C, PT.

96. Gray, *American Civil Liberties,* chapter 7; *BN,* 20–21 April 1934. "Complaints about Japanese growers most numerous": Imperial Valley—May 1934, p. 6, folder 30, carton 14, MSS 84/38C, PT. Glassford's quote: "TO ALL MEXICAN WORKERS IN IMPERIAL VALLEY," General P. D. Glassford, 30 April 1934, folder 11, carton 46, MSS 84/38C, PT. Neutral observer revealed few communist leaders and union's struggle to cultivate leaders: Mr. Cantwell, the editor of the *Brawley Democrat* said, "Vigilantes—at night beat up Mexican leaders who developed. There were only a half dozen communists." CERA Field Notes, 15 June–12 July 1935, folder 39, carton 14, MSS 84/38C, PT.

97. Gray, *American Civil Liberties,* 80–84, 88–90. Wirin's requested investigation of police brutality and violations of civil rights: *BN,* 26–28 April 1934; 2–3 May 1934. Glassford's critical report: ibid., 5 May 1934; Gray, *American Civil Liberties,* 89–90, 100. Glassford's turn against agribusiness: *BN,* 7 May 1934; 9 May 1934; Gray, *American Civil Liberties,* chapter 8. Glassford's views of Terrazas: Imperial Valley—May 1934, p. 6, folder 30, carton 14, MSS 84/38C, PT.

98. Mexican men arrested: *BN,* 8 May 1934; 10 May 1934. Spies in Los Angeles tracking CAWIU leaders: ibid., 22 May 1934; 29 May 1934. Repression: ibid., 18 May 1934; 22 May 1934; 29 May 1934; 1 June 1934. ACLU's request for Glassford's removal: ibid., 15 May 1934; 17 May 1934; Gray, *American Civil Liberties,* 102.

99. *BN,* 5 June 1934; 8–9 June 1934; 11–13 June 1934; Gray, *American Civil Liberties,* 118–20, 135. Scathing press release: *BN,* 14 June 1934. Death threat: ibid., 15 June 1934. A week later, Consul Terrazas and his union's leaders feasted with farm representatives to celebrate the CAWIU's defeat and laud each other for harmonious cooperation: ibid., 22 June 1934. "Communist hysteria": ibid., 26 June 1934. Assault of CAWIU leader: ibid., 27 July 1934; Levenson, "California Casualty List."

100. Terrazas's labor agreement: *CC,* 18 October 1934; *BN,* 18 October 1934. President Rodríguez in Mexicali: *CC,* 5 November 1934; 8 November 1934; 10 November 1934. Terrazas's cantaloupe wage scale: ibid., 27 November 1934.

101. Terrazas's transfer: ibid., 1 February 1935. Terrazas's career and family: ibid., 27 February 1935. Six thousand Filipinos: ibid., 20 November 1935. Bypassing Imperial Valley and demanding land: Andrés, "Invisible Borders," 13.

102. Mexicali strike: *CC,* 28–29 August 1934. Cárdenas visited colonias: ibid., 19 September 1934. 1935 colonia statistics: Sánchez Ramírez, *Crónica agrícola,* 89–90. Cárdenas's objectives to Mexicanize Mexicali: Kerig, "Yankee Enclave," 333–35. Example of union-employer negotiated cotton-picking wage and Mexicali arbitration board approving the agreement: *CC,* 10 September 1936; 15 September 1936; 3 October 1936.

103. *Report Submitted by General Juan Andrew Almaz[á]n, Secretary of Communications and Public Works, to the President of the Republic, with Reference to His Journey through the Northwestern Part of the Country, Mexico, July, 1930,* 29–37, folder Mexico, file 1, IID.

104. Mexicanization: *NYT,* 21 September 1930. Canceling CRLC's concession: *BN,* 30 September

1930. Annexation resolution: *NYT,* 6–7 January 1931. Rumors of annexation and commission's instructions: ibid., 11 January 1931. Cárdenas's water policy and Mexicanization: Dwyer, *Agrarian Dispute,* chapters 2–3. Mexico protested another annexation proposal: *CC,* 1 June 1936.

105. Andrés, "Invisible Borders," 15–18.

106. Asalto: Acosta Montoya, "Precursores del agrarismo," 73–77, 87–109, 129–37; Garduño, *La disputa,* 106–17.

107. Mexicali's ejido movement: Sánchez Ramírez, *Crónica agrícola,* 104–17; Acosta Montoya, "Precursores del agrarismo," 87–139; Garduño, *La disputa,* 117–55. Cárdenas's nationalization land program: Hart, *Empire and Revolution,* chapter 12; Fallaw, *Cárdenas Compromised;* Bantjes, *As If Jesus Walked,* chapter 8; Craig, *First Agraristas,* chapters 4–7; Sanderson, *Agrarian Populism,* chapter 5.

108. Sit-down strike, land seizures, expropriation: Andrés, "Invisible Borders," 19. National media coverage: *NYT,* 23 March 1937; 30 April 1937; 6 May 1937; 13 May 1937.

109. Eyewitness to ejido colonization: Linholm, "A Study of the Agrarian Revolution of Mexico," chapter 5. November statistics: ibid., 63–64. Credit statistics: ibid., 79–80. President Cárdenas's secretly received funding for cotton industry from Clayton Anderson, a wealthy American: Walsh, *Building the Borderlands,* chapter 6. 1937–45 ejido, colonias, acreage statistics: Anguiano Téllez, *Agricultura y migración,* 99–100.

110. CRLC's financial particulars: Stowe, "Pioneering Land." CRLC's viewpoint: Kerig, "Yankee Enclave," 357–98. Mexico's skillful diplomacy over land expropriation repayment: Dwyer, *Agrarian Dispute,* chapters 6–8.

111. Earlier shed strikes: *BN,* 1–2 May 1931; 4–5 May 1931, 13–15 May 1931, 26 June 1931; 2 July 1931; 18 July 1931, 19–20 April 1932; 22 April 1932, 26 April 1932; 12 May 1932; 19–21 May 1932; 23–24 May 1932, 26 May 1932; 3 June 1932. 1935 FVWU strike: ibid., 26 January 1935; 28–29 January 1935; 1 February 1935.

112. *BN,* 13 February 1935. Violence against replacement workers and quote urging fieldworkers to join shed strike: ibid., 14 February 1935. "Feminine strikers": ibid., 15 February 1935. Strike: ibid., 16 February 1935. Deaths of two men: ibid., 18 February 1935. Strike news: ibid., 19–20 February 1935. Coroner's jury recommending grand jury investigation into deaths, "substantial" men; "strike is lost": ibid., 21 February 1935. Chester Moore of Western Growers blamed communists from Los Angeles for the strike: ibid., 22 February 1935. FVWU called off strike: ibid., 25–26 February 1935. Grand jury no charges of murder: ibid., 7 March 1935.

113. Ibid., 26 February 1935; 28 February 1935. Nonstrike bonus: ibid., 8 May 1935. Shippers putting replacement worker as head of Wrap Pack union: ibid., 20 February 1939. Lost wildcat strikes: ibid., 9 December 1935; 30–31 December 1935; 5 February 1936; 14 January 1939. Lost shed strikes: ibid., 5 June 1937; 7 June 1937; 9 June 1937; 7 February 1938. Dueling shed unions: ibid., 7 June 1938; 10 June 1938; 9 December 1938; 19 January 1939; 18 February 1939; 21–22 February 1939; 16 March 1939; 18 March 1939; 24 March 1939; 16–17 May 1939; 20 May 1939; 22 May 1939; 19 June 1939; 26 June 1939; 29 June 1939; 18 July 1939. Overview of shed unionization: Ruiz, *Cannery Women;* Stein, *California and the Dust Bowl,* chapters 8–9.

114. Mini, "California Dictatorship."

115. "Voices from the Dustbowl."

Conclusion

1. *BN,* 6 November 1930.

2. Ibid.

3. McWilliams, *California,* 162.

4. John Phillips, C. B. Hutchinson, W. C. Jacobsen, *The Imperial Valley Farm Labor Situation, 16 April 1934,* p. 14–16, folder 11, carton 46, MSS 84/38C, PT.

Afterword

1. Imperial County, California. http://quickfacts.census.gov/qfd/states/06/06025.html; *IVP,* 30 April 2008; 24 July 2008; 29 September 2009; 1 January 2010; 6 October 2009; *NYT,* 16 March 2011; *IVP,* 9 January 2010; 1 December 2011; 10 January 2009.

2. *DS,* 1 June 2012; *IVP,* 4 May 2013.

3. *IVP,* 18 November 2009; 28 November 2009; 10 November 2011; 30 June 2013; *LAT,* 16 July 2012; *IVP,* 10 August 2013; Law and Resource, *Critical Path Forward,* 96–100.

4. Ward, *Border Oasis,* 37–38.

5. Directorio de la Industria Maquiladora de Baja California: *IVP,* 16 January 2013; *LAT,* 26 May 2012.

6. Fradkin, *River No More,* 276–87; Kinsley, "Ben Yellen's"; Yellen, "Mimeograph Politics"; Taylor, "Mexican Migration and 160-Acre Water Limitation," 745–47. Overview: Taylor, "Water, Land, and Environment [in] Imperial Valley"; Koppes, "Public Water, Private Land"; Kelley, "Reclamation Law in Litigation."

7. Nathanson, *Updating Hoover Dam,* appendix X, x–68. Present perfect rights: ibid., chapter 10. 1979 Supreme Court supplemental decree: US Department of the Interior, Bureau of Reclamation, *Reclamation, Managing Water in the West,* Supplement 4, p. 429.

8. *Bryant v. Yellen,* 447 U.S. 352 (1980); US Department of the Interior, Bureau of Reclamation, *Reclamation, Managing Water in the West,* 12–9 to 12–13; Hundley, *Great Thirst,* 223–26.

9. *NYT,* 23 October 2011; *Imperial County, California, Annual Agricultural Crop and Livestock Report, 1929.*

10. Law and Resource, *Critical Path Forward,* 58.

11. US War Relocation Authority, *West Coast Localities;* Overview: Loveridge-Sanbonmatsu, *Imperial Valley Nisei Women.* Sanbonmatsu family successful farmers after World War II: *IVP,* 5 August 2010.

12. *Imperial County, Annual Agricultural Crop and Livestock Reports, 2003–11.* Beef plant closure: *IVP* 31 January 2014; 2 February 2014; 4–5 February 2014; 9 February 2014; 22–23 February 2014.

13. Total ($112,111,698) for six zip codes Imperial County (92227, 92233, 92250, 92243, 92251, 92257) 1995–2012, Environmental Working Group, Farm Subsidy Database.

14. *Imperial County, Agricultural Crop and Livestock Reports, 1949–50.*

15. Ibid., 2007–11.

16. *SB,* 28 April 2012; 30 April 2012.

17. Bracero program in California: Galarza, *Farm Workers and Agri-Business in California;* Mitchell, *They Saved the Crops.* Overview in Imperial Valley: Galarza, *Farm Workers and Agri-Business in California,* 146–86; Mitchell, *They Saved the Crops,* chapter 6; Wolf, "The Imperial Valley as an Index of Agricultural Labor Relations," chapter 10. Wages: Lawson, "Bracero," 66–67, 80. 10,000: Ward, *Border Oasis,* 29.

18. Bracero Program overview: Galarza, *Merchants of Labor.* Imperial Valley: Lawson, "Bracero," 53, 55. 1960–61 strike: Mitchell, *They Saved the Crops,* 355–60. Labor abuses and 1960–61 strike: Wolf, "The Imperial Valley as an Index of Agricultural Labor Relations," chapter 11; Cohen, *Braceros,* 8, 139–40, 156–58, 164. UFW's defeated strike: *LAT,* 14 May 1986. No current field union representation: Martin, "Imperial Valley."

19. Taylor, "Immigration, Employment, Poverty, and Welfare"; *Imperial County, Agricultural Crop and Livestock Reports, 2003–2006; IVP,* 3 December 2011; Vollmann, *Imperial,* 358–61; *IVP,* 16 January 2013.

20. Hundley, *Great Thirst,* 473–78.

21. Ibid., 478–500.

22. US Department of the Interior, Bureau of Reclamation, *Reclamation, Managing Water in the West,* 6–9 to 6–11; *NYT,* 11 December 2002; 12 January 2003; 7 September 2003; 17 October 2003; *SD,* 17 December 2009; *Quantification Settlement Agreement Cases,* JCCP 4353, December 7, 2011; *IVP,* 5 June 2013.

23. *SUN,* 21 April 2013; *DS,* 22 September 2012; *IVP,* 12 February 2010; *SD,* 18 November 2011; Dowd, "Imperial Valley, Salton Sea and Delta," CL; Cohen, *Hazard,* iii, 4, 36.

24. Cohen, *Hazard,* 33–35; *IVP,* 24 January 2009; 26 January 2009; 20 October 2009; 28 January 2010; *SD,* 15 January 2010. Brad Poiriez: *IVP,* 5 June 2013. Tim Krantz: *SD,* 18 November 2011.

25. California Legislative Analyst's Office. Army Corp of Engineers mitigate project: *SUN,* 21 April 2013. Kevin Kelley: *SD,* 5 August 2013; May 7, 2013. Imperial Irrigation District and Imperial County term sheet.

26. *IVP,* 26 July 2008; 11 December 2009; 9 October 2012; QSA Coordinated Civil Cases; *IVP,* 1 August 2013.

27. *IVP,* 1 April 2009; 10 August 2011; *DS,* 22 December 2011; Law and Resource, *Critical Path Forward,* 103–7. Fallowing program paying farmers not to farm: Imperial Irrigation District board meeting agenda August 20, 2013, pp. 036–039. 2014 fallowing program pays $150 per acre-foot with the tenant receiving 20 percent and the landowner pocketing 80 percent: *IVP,* 26 March 2014. Few farm water efficiency contracts: *IVP,* 5 December 2013; Than, "Can California Farmers Save Water."

28. *IVP,* 30 December 2009; 1 January 2010; 25 February 2011; 6 April 2011; 12 April 2011; 29 May 2013.

29. Ward, *Border Oasis,* chapters 3–6.

30. Bergman, *Red Delta,* chapter 7.

31. *NYT,* 15 April 2013; *LAT,* 24 May 2008.

32. Tisdale, "Cocopah Identity," 150–54, 161–64, 172–76; Williams, "People and River"; Williams, *Cocopah People; LAT,* 25 March 2012; Bergman, *Red Delta,* chapters 9, 11–13.

33. *LV,* 8 November 2012; *LAT,* 20 November 2012; 26 November 2012; *NYT,* 15 April 2013. The "pulse flow" invigorates the delta: *IVP,* 12 March 2014; *LAT,* 23 March 2014; *IVP,* 28 March 2014.

34. *LAT,* 13 December 2012; *SB,* 16 April 2013; *LAT,* 28 May 2013; US Department of Interior, *Colorado River Basin Water Supply and Demand Study,* 3, 8–9. Tina Shields: *LAT,* 27 May 2013. The drought highlights the distorted desert agribusiness economy. The majority of the Imperial Valley's subsidized water irrigates low value forage crops like alfalfa for export to Japan, the United Arab Emirates, and China to feed dairy cows: Jervey, "Exporting the Colorado River to Asia, Through Hay." High water consumption crops permit farmers to inflate their water use history, which reaps larger payments for selling water rather than growing food. The Imperial Irrigation District recently apportioned all of the farm water to individual fields, a landmark decision that ties water to land for the first time in the valley's history. Field water apportionment is the latest step in transferring the Colorado River into a private water right. Moreover, the IID created a clearinghouse so farmers can sell water to each other without reporting the transaction's financial particulars to the district: *IVP,* 28 October 2013.

Bibliography

Primary Sources

Archival Collections

BL	Bancroft Library. University of California, Berkeley.
BPL	Brawley Public Library.
CL	Imperial Valley Records. Water Resources Collection. Special Collections. Honnold/ Mudd Library. The Claremont Colleges. Claremont, California.
FA	Inventory of the Frank Adams Papers, 1889–1962. From the Holdings of the Special Collections and Archives. Water Resources Collections. UCR Libraries. University of California, Riverside.
HH	Herbert Hoover Collection. Hoover Institution Archives. Stanford University.
HL	Huntington Library. San Marino, California.
IID	Imperial Irrigation District Archive. Imperial, California.
PM	Imperial County Historical Society. Pioneers Museum. Imperial, California.
PT	Paul S. Taylor Papers. Bancroft Library. University of California, Berkeley.
RG 25	Record Group 25. Records of the National Labor Relations Board District 15 (Los Angeles). National Archives and Records Administration, Pacific Region. Perris, California.
RG 85	Record Group 85. Records of the Immigration and Naturalization Service Series A. Subject Correspondence Files pt. 2: Mexican Immigration 1906–30. Microfilmed by University Publications of America. Bethesda, Maryland, 1993.
SL	Colorado River Land Company. Sociedad Anónima. Sherman Library. Corona del Mar, California.
SRR	Survey of Race Relations Collection. Hoover Institution Archives. Stanford University.

Newspapers

BN	*Brawley News*, 1900–39
CC	*Calexico Chronicle*, 1919–30, 1934–39
DS	*Desert Sun*
ECP	*El Centro Progress*
IVP	*Imperial Valley Press*, 2002–13
LAT	*Los Angeles Times*
LV	*Las Vegas Review-Journal*
SB	*Sacramento Bee*
SD	*San Diego Union-Tribune*
SUN	*San Bernardino Sun*
NYT	*New York Times*

Miscellaneous

Adams, Frank. *Irrigation from Colorado River in Lower California*. Department of State. International Water Commission United States and Mexico. June 1928. http://cdm16658.contentdm.oclc.org/cdm/ref/collection/p267501ccp2/id/1808.

———. *Progress Report of Special Colorado River Investigations, June, 1926, to April, 1927*. US Department of the Interior. Bureau of Reclamation. May 10, 1927. http://cdm16658.contentdm.oclc.org/cdm/ref/collection/p267501ccp2/id/1784.

———. *Report of Inspection Trip over Colorado River Levee Systems below Yuma, Arizona, September 17, 18, 19, 1929*. International Water Commission United States and Mexico. http://cdm16658.contentdm.oclc.org/cdm/ref/collection/p267501ccp2/id/1812.

Anderholt, Joseph J. *Desert Dairies: Catalyst for the Development of Imperial Valley*. 3rd ed. Imperial, CA: Imperial County Historical Society, 1991.

Anderholt, Joseph J., and Dorothy M. Anderholt, eds. *The History of the Imperial Valley Swiss*. Holtville, CA: Imperial Valley Swiss Club, 1984.

Answer to Brockman-Beal Letter as Given by One Imperial Valley Landowner. Imperial, CA: Imperial Irrigation District, July 28, 1931. CL.

Barnett, Paul. *Imperial Valley: The Land of Sun and Subsidies*. Davis: California Institute for Rural Studies, 1978.

Bryant v. Yellen. 447 U.S. 352 (1980). https://bulk.resource.org/courts.gov/c/US/447/447.US.352.79-435.79-425.79-421.html.

Burgess, Charles P. *Imperial Irrigation District, California*. Blyth and Co. and Kaiser and Co., 1943.

California Legislative Analyst's Office. http://www.lao.ca.gov/2008/rsrc/salton_sea/salton_sea_01-24-08.aspx.

Cohen, Michael J., and Karen H. Hyun. *Hazard: The Future of the Salton Sea with No Restoration Project*. Pacific Institute, May 2006.

Cramp, Kathryn, Louise F. Shields, and Charles A. Thomson. *Study of the Mexican Population in Imperial Valley, California*. For the Committee on Farm and Cannery Migrants, Council of Women for Home Missions, New York, 31 March–9 April 1926. BL.

de Vivanco, Aurelio. *Baja California al Dia, Lower California up to Date*. 1924.

Directorio de la Industria Maquiladora de Baja California (August 5, 2013).

Dowd, M. J. *History of Imperial Irrigation District and the Development of Imperial Valley*. Imperial Irrigation District, 1956.

———. "Imperial Valley, Salton Sea, and the Delta of the Colorado River." Address to the American Water Works Assn. Palm Springs, CA. April 24, 1953. CL.

Environmental Working Group, Farm Subsidy Database, 1995–2012. http://farm.ewg.org/index.php?key=nosign.

Frisby, Karen J. *Imperial Valley: The Greening of a Desert*. Imperial County, CA: Imperial Valley College Desert Museum Society, 1993.

Heffernan, W. T. *Personal Recollections of the Early History of Imperial Valley*. Imperial Valley Pioneers Association, 1928.

Holt, L. M., compiler. *The Imperial Valley Scrapbook: Containing Clippings from the Press of the Country Relative to the Reclamation and Settlement of the Imperial Valley, with Occasional Introductions, Notes, and Comments by the Compiler*. Los Angeles: Southern California Printing, n.d. CL.

Howe, E. F. "Calexico, Imperial Valley: The Delta of the Colorado, An Inland Empire of Richness." In *Calexico, California: "King Cotton's Capital,"* 8–14. The Farmers and Merchants Club and City of Calexico, 1915.

Imperial County, California. http://quickfacts.census.gov/qfd/states/06/06025.html.

Imperial Irrigation District and Imperial County term sheet, May 7, 2013. http://www.iid.com/modules/showdocument.aspx?documentid=7818.

Imperial Land Company. *From Desert to Garden, from Worthlessness to Wealth as Illustrated in the Imperial Settlements: San Diego County, Southern California*. Times-Mirror Printing and Binding House, Los Angeles, May 1902. CL.

———. *In a Nutshell*. Circa 1901. CL.

Imperial Valley, California. San Francisco: Passenger Department Southern Pacific, 1908.

Imperial Valley's Most Essential Need Is a Flood Control and Storage Dam in the Colorado River: The Proposed All-American Canal Is a Non-Essential, 1924. HL.

Johnson, Hiram W. "The Epic of the Imperial Valley." An Address to the Commonwealth Club of San Francisco, October 17, 1925. CL.

Kelly, Allen. "Calexico: King Cotton's Capital." In *Calexico, California: "King Cotton's Capital,"* 1–7. The Farmers and Merchants Club and City of Calexico, 1915.

Kelley, Amy K. "Reclamation Law in Litigation: Acreage and Residency Limitations on Private Lands." 1976. http://nationalaglawcenter.org/publication/kelley-acreage-and-residency-limitations-in-the-imperial-valley-a-case-study-in-national-reclamation-policy-23-south-dakota-l-rev-621-661-1978/wppa_open/.

Law and Resource Planning Associates. *A Critical Path Forward: A QSA Plan B for Protecting Water Rights, the Environment, and the People of the Imperial Valley.* September 2012.

Martin, Philip. "Imperial Valley: Agriculture and Farm Labor." *Changing Face* 7, no. 1 (2001). http://migration.ucdavis.edu/cf/comments.php?id=34_0_2_0.

McNaught, Margaret S. "The Squirrel Is a Hun." *California Blue Bulletin* 4, no. 2 (1918): 16.

Ockerson, J. A. *Report on Lower Colorado River and Delta.* October 4, 1910. http://archive.org/details/reportonlowercolooockerich.

Quantification Settlement Agreement Cases. JCCP 4353. December 7, 2011.

QSA Coordinated Civil Cases. JCCP 4353. July 31, 2013.

Spector, Frank. *Free the Imperial Valley Prisoners.* International Labor Defense, March 1932.

———. *Story of the Imperial Valley.* International Labor Defense, n.d.

Taylor, Edward J. "Immigration, Employment, Poverty, and Welfare in the Imperial Valley." *Changing Face* 7, no 1 (2001). http://migration.ucdavis.edu/cf/more.php?id=35_0_2_0.

Tout, Otis B., ed. *The First Thirty Years, 1901–1931.* San Diego: Arts and Crafts Press, 1931; rpt. Imperial County Historical Society, 1990.

"Voices from the Dustbowl." National Public Radio, All Things Considered, Lost and Found Sound, July 28, 2000. Videocassette.

The Winning of Barbara Worth. Produced by Samuel Goldwyn. Goldwyn Productions, 1926. Videocassette.

Census Records

US Department of Commerce. Bureau of the Census. *Census of Agriculture: 1945* vol. 1, pt. 33, *California: Statistics for Counties.* Washington, DC: Government Printing Office, 1946.

———. *Fifteenth Census of the United States: 1930* vol. 2, pt. 3, *Agriculture.* Washington, DC: Government Printing Office, 1932.

———. *Fifteenth Census of the United States: 1930* vol. 3, pt. 1, *Population.* Washington, DC: Government Printing Office, 1932.

———. *Fifteenth Census of the United States: 1930. Irrigation of Agricultural Lands.* Washington, DC: Government Printing Office, 1932.

———. *Sixteenth Census of the United States: 1940* vol. 2, pt. 1, *Population.* Washington, DC: Government Printing Office, 1943.

———. *Sixteenth Census of the United States: 1940* vol. 2, pt. 3, *Agriculture: Statistics for Counties.* Washington, DC: Government Printing Office, 1942.

———. *Statistical Atlas of the United States, 1914.* Washington, DC: Government Printing Office, 1914.

———. *Thirteenth Census of the United States: Taken in the Year 1910, Abstract of the Census, Supplement for California.* Washington, DC: Government Printing Office, 1913.

Government Publications

Bell, P. L., and H. Bentley Mackenzie. *Mexican West Coast and Lower California: A Commercial and Industrial Survey.* Department of Commerce. Washington, DC: Government Printing Office, 1923.

Bloch, Louis. *Report on the Mexican Labor Situation in the Imperial Valley.* 22nd Biennial Report of the Bureau of Labor Statistics of the State of California, 1925–1926. Sacramento: California Bureau of Labor Statistics, 1926.

Decisions of the Department of the Interior and General Land Office in Cases Relating to the Public Lands from June 1, 1904, to June 30, 1905. 33. Washington, DC: Government Printing Office, 1905.

Dixon, Joseph. *Control of the Pocket Gopher in California.* College of Agriculture, Agricultural Experiment Station Bulletin 340. Berkeley: University of California Press, 1922.

Imperial County, California. Annual Agricultural Crop and Livestock Reports, 1907–39, 1949–50, 2003–11. http://www.co.imperial.ca.us/ag/Departments_A/agricultural_crop_&_livestock_reports.htm.

McLachlan, Argyle. *Community Production of Durango Cotton in the Imperial Valley*. 324. United States Department of Agriculture. Washington, DC: Government Printing Office, December 22, 1915.

Mexicans in California: Report of Governor C. C. Young's Mexican Fact Finding Committee. Sacramento: California State Printing Office, 1930; rpt. San Francisco: R. and E. Associates, 1970.

Nathanson, Milton N. *Updating the Hoover Dam Documents*. Washington DC: Government Printing Office, 1980. In *Reclamation, Managing Water in the West: The Colorado River Documents 2008*. DVD. US Department of the Interior. Bureau of Reclamation. September 2010.

State Board of Control of California. *California and the Oriental: Japanese, Chinese, and Hindus*. Sacramento: California State Printing Office, 1922.

Tait, C. E. *Irrigation in Imperial Valley, California: Its Problems and Possibilities*. Washington, DC: Government Printing Office, 1908.

US Congress. House. Committee on Foreign Affairs. *Report of the American Section of the International Water Commission United States and Mexico*. H.R. 359. 71st Cong., 2nd sess. Washington, DC: Government Printing Office, 1930.

US Congress. Senate. Committee on Irrigation and Reclamation. *Boulder Canyon Reclamation Project*. Report 654. Part 1. S. 3331. 69th Cong. 1st sess. Washington, DC: Government Printing Office, 1926.

US Department of the Interior. Bureau of Reclamation. *Boulder Canyon Project All-American Canal: Contract for Construction of Diversion Dam, Main Canal and Appurtenant Structures and for Delivery of Water and Contract Amendatory of and Supplemental to All-American Canal Contract of December 1, 1932*. Washington, DC: Government Printing Office, March 4, 1952.

———. *Colorado River Basin Water Supply and Demand Study: Final Study Reports*. December 2012.

———. *Reclamation, Managing Water in the West: The Colorado River Documents 2008*. September 2010. DVD.

US Department of the Interior. Reclamation Service. *Preliminary Report on Problems of Imperial Valley and Vicinity*. Washington, DC, January 1921.

US War Relocation Authority. Community Analysis Section. 12. *West Coast Localities: Imperial Valley*. Washington, DC, April 9, 1945.

Congressional Hearings

US Congress. House. Committee on Immigration and Naturalization. *Immigration from Countries of the Western Hemisphere*. Hearings on H.R. 6465. H.R. 7358. H.R. 10955. H.R. 11687. 70th Cong., 1st sess. Washington, DC: Government Printing Office, 1928. http://babel.hathitrust.org/cgi/pt?id=mdp.39015014316957;page=root;view=image;size=100;seq=1.

US Congress. House. Committee on Immigration and Naturalization. *Western Hemisphere Immigration*. Hearings on H.R. 8523. H.R. 8530. H.R. 8702. 71st Cong., 2nd sess. Washington, DC: Government Printing Office, 1930. http://babel.hathitrust.org/cgi/pt?id=mdp.39015014316932.

US Congress. House. Committee on Irrigation of Arid Lands. *All-American Canal in Imperial and Coachella Valleys, Calif.* Hearings before the Committee on Irrigation of Public Lands. House of Representatives. 66th Cong. Washington, DC: Government Printing Office, 1920. http://books.google.com/books?id=19YmJGCbuIIC&printsec=frontcover&source=gbs_ge_summary_r&cad=0#v=onepage&q&f=false.

US Congress. House. Committee on Irrigation and Reclamation. *Protection and Development of Lower Colorado River Basin*. Information on H.R. 2903. 68th Cong., 1st sess. Washington, DC: Government Printing Office, 1924.

US Congress. Senate. Committee on Immigration. *Restriction of Western Hemisphere Immigration*. H.R. 1437. 70th Cong., 1st sess. Washington, DC: Government Printing Office, 1928. http://babel.hathitrust.org/cgi/pt?id=mdp.39015014316940.

US Congress. Senate Subcommittee of the Committee on Education and Labor. *Violations of Free*

Speech and Rights of Labor. Hearings Pursuant to Senate. Res. 266 76th Cong. Washington, DC: Government Printing Office, 1940.

Theses and Dissertations

Aguila, Jaime R. "Protecting 'Mexico de Afuera': Mexican Emigration Policy, 1876–1928." PhD diss., Arizona State University, 2008. ProQuest (AAT 9990715).

Andrés, Benny Joseph, Jr. "Power and Control in Imperial Valley, California: Nature, Agribusiness, Labor, and Race Relations, 1900–1940." PhD diss., University of New Mexico, 2003.

Elliott, Robert S. "The Health and Relief Problems of a Group of Non-Family Mexican Men in Imperial County, California." Master's thesis, University of Southern California, 1939.

Fuller, Levi V. "The Supply of Agricultural Labor as a Factor in the Evolution of Farm Organization in California." PhD diss., University of California, Berkeley, 1939.

Hendricks, William O. "Guillermo Andrade and Land Development on the Mexican Colorado River Delta, 1878–1905." PhD diss., University of Southern California, 1967.

Holliday, Jay N. "A Study of Non-Attendance in Miguel Hidalgo School of Brawley, California." Master's thesis, University of Southern California, 1935.

Kerig, Dorothy P. "Yankee Enclave: The Colorado River Land Company and Mexican Agrarian Reform in Baja California, 1902–1944." PhD diss., University of California, Irvine, 1988.

Kim, Mee-Ae. "Immigrants, Workers, Pioneers: The Chinese and Mexican Colonization Efforts, 1890–1930." PhD diss., Washington State University, 2000.

Lawson, Luther W. "The Bracero in California's Imperial Valley." Master's thesis, University of Southern California, 1965.

Linholm, Earl. "A Study of the Agrarian Revolution of Mexico as Applied to the Division of Land among the Peons in the Mexicali Region of Lower California." Master's thesis, University of Southern California, 1938.

Peters, James P. "Agricultural Unionism in the Imperial Valley, 1928–1940." Master's thesis, San Diego State University, 1967.

Tisdale, Shelby Jo-Anne. "Cocopah Identity and Cultural Survival: Indian Gaming and the Political Ecology of the Lower Colorado River Delta, 1850–1996." PhD diss., University of Arizona, 1997.

Walton, Roger M. V. "A Study of Migratory Mexican Pea-Pickers in Imperial Valley." Master's thesis, University of Southern California, 1940.

Wolf, Jerome. "The Imperial Valley as an Index of Agricultural Labor Relations in California." PhD diss., University of Southern California, 1964.

Interviews

Chikahisa, Ray. Interview by author, May 11, 2001. Tape recording. Imperial.

Gongora, Delia. Interview by author, November 24, 1999 and February 28, 2000. Tape recording. El Centro.

Ibarra, Modesto. Interview by author, January 15, 1998. Tape recording. Brawley.

Laughrin, Richard. Interview by author, January 14, 1998. Tape recording. Brawley.

Nimura, Takanori (Pro). Interview by author, March 18 and 19, 1998. Tape recording. Brawley and Imperial.

Sano, Peter Iwao. Interview by author, August 13, 1999. Tape recording. Palo Alto.

Taylor, Kathryn. Interview by author, July 26, 1999. Tape recording. Calipatria.

Thornburg, William. Interview by author, January 13, 1998. Tape recording. Holtville.

Zinn, Evelyn. Interview by author, March 5, 1999. Tape recording. El Centro.

Journals and Magazines

Carr, Harry. "The Kingdom of Cantu: Why Lower California Is an Oasis of Perfect Peace in Bloody Mexico." *Sunset* 38 (1917): 33–34, 65–67.

Duryea, Edwin, Jr. "The Salton Sea Menace." *Out West* 24, no. 1 (1906): 3–24. CL.

Hall, Wilbur J. "Just Like Dixie Land." *Sunset Magazine* 24, no. 2 (1910): 175.

Holt, L. M. "How the Reclamation Service Is Robbing the Settler." *Overland Monthly* (November 1907): 510–12.

———. "The Reclamation Service and the Imperial Valley." *Overland Monthly* (January 1908): 70–75.

Hosmer, Helen. "Who Are the Associated Farmers?" *Rural Observer* 1, no. 9 (1938): 2–19.

"Imperial Valley." *Calexico Chronicle*. 2nd Annual Magazine Edition. May 1909. CL.

Jervey, Ben. "Exporting the Colorado River to Asia, through Hay." *National Geographic* (January 23, 2014). http://news.nationalgeographic.com/news/2014/01/140123-colorado-river-water-alfalfa-hay-farming-export-asia/.

Jones, Ellis O. "Kidnap Valley, California." *Nation* (April 25, 1934): 468–70.

Kenamore, Clair. "The Principality of Cantu." *Bookman* 61 (1917): 23–28.

Kinsley, Michael E. "Ben Yellen's Fine Madness." *Washington Monthly* (January 1971): 38–49.

Levenson, Lew. "California Casualty List." *Nation* (August 29, 1934): 243–45.

Mini, Norman. "That California Dictatorship." *Nation* (February 20, 1935): 224–26.

Mitchell, Edmund. "The Salton Sea." *North American Review* 183, no. 597 (1906): 224–36. http://www.jstor.org/stable/25105611.

Rush, Emmy M. "Imperial Valley, the Wonderful." *Overland Monthly* 78 (December 1921): 54–56, 67–69.

Than, Ker. "Can California Farmers Save Water and the Dying Salton Sea?" *National Geographic* (February 18, 2014). http://news.nationalgeographic.com/news/2014/02/140218-salton-sea-imperial-valley-qsa-water-conservation/.

Williams, Chester S. "Imperial Valley Mob." *New Republic* 78 (February 21, 1934): 39–41.

———. "Imperial Valley Prepares for War." *World Tomorrow* 17 (April 26, 1934): 199–201.

Yellen, Ben. "The Mimeograph Politics of Dr. Ben Yellen: The Elmore Family." *California Homeowner Magazine* 12, no. 2 (1973): 18–20.

Secondary Sources

Acosta Montoya, David. *"Precursores del agrarismo" y "El asalto a las tierras" en el estado de Baja California*. Mexicali, Baja California: Instituto de Investigaciones Históricas, 1985.

Aguirre Bernal, Celso. *Compendio histórico-biográfico de Mexicali, 1539–1966*. 7th ed. Mexicali, Baja California, 1966.

Alamillo, José M. *Making Lemonade Out of Lemons: Mexican American Labor and Leisure in a California Town, 1880–1960*. Urbana: University of Illinois Press, 2006.

Alanís Enciso, Fernando Saúl. *Que se queden allá: El gobierno de México y la repatriación de mexicanos en Estados Unidos, 1934–1940*. Tijuana, Baja California: El Colegio de la Frontera Norte, 2007.

Alexander, J. A. *The Life of George Chaffey: A Story of Irrigation Beginnings in California and Australia*. Melbourne: Macmillan, 1928.

Almaguer, Tomás. *Racial Fault Lines: The Historical Origins of White Supremacy in California*. Berkeley: University of California Press, 1994.

Alvarez, Robert R., Jr. *Familia: Migration and Adaptation in Baja and Alta California, 1800–1975*. Berkeley: University of California Press, 1991.

Andrews, Thomas G. *Killing for Coal: America's Deadliest Labor War*. Cambridge, MA: Harvard University Press, 2008.

Andrés, Benny J., Jr. "'I Am Almost More at Home with Brown Faces Than with White': An Americanization Teacher in Imperial Valley, California, 1923–1924." *Southern California Quarterly* 93, no. 1 (2011): 69–107.

———. "Invisible Borders: Repatriation and Colonization of Mexican Migrant Workers along the California Borderlands during the 1930s." *California History* 88, no. 4 (2011): 5–21, 63–65.

Anguiano Téllez, María Eugenia. *Agricultura y migración en el Valle de Mexicali*. Tijuana, Baja California: El Colegio de la Frontera Norte, 1995.

Azuma, Eiichiro. *Between Two Empires: Race, History, and Transnationalism in Japanese America*. New York: Oxford University Press, 2005.

Balderrama, Francisco E., and Raymond Rodríguez. *Decade of Betrayal: Mexican Repatriation in the 1930s*. Rev. ed. Albuquerque: University of New Mexico Press, 2006.

Baldoz, Rick. *The Third Asiatic Invasion: Empire and Migration in Filipino America, 1898–1946*. New York: New York University Press, 2011.

Bantjes, Adrian A. *As If Jesus Walked on Earth: Cardenismo, Sonora, and the Mexican Revolution*. Wilmington, DE: SR Books, 1998.

Barajas, Frank P. *Curious Unions: Mexican American Workers and Resistance in Oxnard, California, 1898–1961*. Lincoln: University of Nebraska Press, 2012.

Barton Josef. "At the Edge of the Storm: Northern Mexico's Rural Peoples in a New Regime of Consumption, 1880–1940." In *Land of Necessity: Consumer Culture in the United States-Mexico Borderlands*, edited by Alexis McCrossen, 217–47. Durham, NC: Duke University Press, 2009.

———. "Borderland Discontents: Mexican Migration in Regional Contexts, 1880–1930." In *Repositioning North American Migration History: New Directions in Modern Continental Migration, Citizenship, and Community*, edited by Marc S. Rodriguez, 141–205. Rochester, NY: University of Rochester Press, 2004.

Benton-Cohen, Katherine. *Borderline Americans: Racial Division and Labor War in the Arizona Borderlands*. Cambridge, MA: Harvard University Press, 2009.

Bergman, Charles. *Red Delta: Fighting for Life at the End of the Colorado River*. Golden, CO: Fulcrum, 2002.

Blackbourn, David. *The Conquest of Nature: Water, Landscape, and the Making of Modern Germany*. New York: W. W. Norton, 2006.

Blaisdell, Lowell L. *The Desert Revolution. Baja California, 1911*. Madison: University of Wisconsin Press, 1962.

Boime, Eric. "'Beating Plowshares into Swords': The Colorado River Delta, the Yellow Peril, and the Movement for Federal Reclamation, 1901–1928." *Pacific Historical Review* 78, no. 1 (2009): 27–53.

———. "National Moat, Regional Lifeline: The Campaign for the All-American Canal, 1917–1944." *Journal of Borderland Studies* 26, no. 2 (2011): 161–78.

Boyer, Christopher R. *Becoming Campesinos: Politics, Identity, and Agrarian Struggle in Postrevolutionary Michoacán, 1920–1935*. Stanford, CA: Stanford University Press, 2003.

Burke, Robert E. *Olson's New Deal for California*. Berkeley: University of California Press, 1953.

Cadava, Geraldo L. "Borderlands of Modernity and Abandonment: The Lines within Ambos Nogales and the Tohono O'odham Nation." *Journal of American History* 98, no. 2 (2011): 362–83.

Calderón, Roberto R. *Mexican Coal Mining Labor in Texas and Coahuila, 1880–1930*. College Station: Texas A&M University Press, 2000.

Cardoso, Lawrence A. "Labor Emigration to the Southwest, 1916 to 1920: Mexican Attitudes and Policy." *Southwestern Historical Quarterly* 79, no. 4 (1976): 400–416. www.jstor.org/stable/30238403.

———. *Mexican Emigration to the United States, 1897–1931: Socio-Economic Patterns*. Tucson: University of Arizona Press, 1980.

Chamberlin, Eugene K. "Mexican Colonization versus American Interests in Lower California." *Pacific Historical Review* 20, no. 1 (1951): 43–55.

Chambers, Clarke A. *California Farm Organizations: A Historical Study of the Grange, the Farm Bureau, and the Associated Farmers, 1929–1941*. Berkeley: University of California Press, 1952.

Chang, Kornel. *Pacific Connections: The Making of the U.S.-Canadian Borderlands*. Berkeley: University of California Press, 2012.

Chase, J. Smeaton. *California Desert Trails*. Boston: Houghton Mifflin, 1919.

Cohen, Deborah. *Braceros: Migrant Citizens and Transnational Subjects in the Postwar United States and Mexico*. Chapel Hill: University of North Carolina Press, 2011.

Cooper, Erwin. *Aqueduct Empire: A Guide to Water in California, Its Turbulent History and Its Management Today*. Glendale, CA: Arthur H. Clark, 1968.

Cory, H. T., and W. P. Blake. *The Imperial Valley and the Salton Sink*. San Francisco: J. J. Newbegin, 1915.

Craig, Ann L. *The First Agraristas: An Oral History of a Mexican Agrarian Reform Movement*. Berkeley: University of California Press, 1983.

Cronon, William. *Nature's Metropolis: Chicago and the Great West*. New York: W. W. Norton, 1991.

Daniel, Cletus E. *Bitter Harvest: A History of California Farmworkers, 1870–1941*. Ithaca, NY: Cornell University Press, 1981.

deBuys, William, and Joan Myers. *Salt Dreams: Land and Water in Low-Down California*. Albuquerque: University of New Mexico Press, 1999.
Dixon, Joseph. "Rodents and Reclamation in the Imperial Valley." *Journal of Mammalogy* 3, no. 3 (1922): 136–46. http://www.jstor.org/stable/1373657.
Duke, Emma. "California the Golden." In *Children in the Fields*, edited by Dan C. McCurry, 233–56. New York: Arno Press, 1975; rpt. *The American Child*, New York, 1920.
Duncan, Robert H. "The Chinese and the Economic Development of Northern Baja California, 1889–1929." *Hispanic American Historical Review* 74, no. 4 (1994): 615–47.
Dwyer, John J. *The Agrarian Dispute: The Expropriation of American-Owned Rural Land in Postrevolutionary Mexico*. Durham, NC: Duke University Press, 2008.
España-Maram, Linda. *Creating Masculinity in Los Angeles's Little Manila: Working-Class Filipinos and Popular Culture, 1920s–1950s*. New York: Columbia University Press, 2006.
Ettinger, Patrick. *Imaginary Lines: Border Enforcement and the Origins of Undocumented Immigration, 1882–1930*. Austin: University of Texas Press, 2009.
Evans, Sterling. *Bound in Twine: The History and Ecology of the Henequen-Wheat Complex for Mexico and the American and Canadian Plains, 1880–1950*. College Station: Texas A&M University Press, 2007.
———. "Yaquis vs. Yanquis: An Environmental and Historical Comparison of Coping with Aridity in Southern Sonora." *Journal of the Southwest* 40, no. 3 (1998): 363–96.
Fallaw, Ben. *Cárdenas Compromised: The Failure of Reform in Postrevolutionary Yucatán*. Durham, NC: Duke University Press, 2001.
Farr, Fenis C., ed. *The History of Imperial County, California*. Berkeley: Elms and Franks, 1918.
Fisher, Lloyd H. *The Harvest Labor Market in California*. Cambridge, MA: Harvard University Press, 1953.
Fiege, Mark. *Irrigated Eden: The Making of an Agricultural Landscape in the American West*. Seattle: University of Washington Press, 1999.
Foley, Neil. *The White Scourge: Mexicans, Blacks, and Poor Whites in Texas Cotton Culture*. Berkeley: University of California Press, 1997.
Fradkin, Philip L. *A River No More: The Colorado River and the West*. Berkeley: University of California Press, 1996.
Galarza, Ernesto. *Farm Workers and Agri-Business in California, 1947–1960*. Notre Dame, IN: University of Notre Dame Press, 1977.
———. *Merchants of Labor: The Mexican Bracero Story*. Charlotte, NC: McNally and Loftin, 1964.
García, Matt. *A World of Its Own: Race, Labor, and Citrus in the Making of Greater Los Angeles, 1900–1970*. Chapel Hill: University of North Carolina Press, 2001.
Garduño, Everardo. *La disputa por la tierra . . . La disputa por la voz: Historia oral del movimiento agrario en el valle de Mexicali*. Mexicali, Baja California: Universidad Autónoma de Baja California, 2004.
Gerardo, Eduardo A. *El dragón en el desierto: Los pioneros chinos en Mexicali*. Mexicali, Baja California: Instituto de Cultura de Baja California, 1991.
Gerhard, Peter. "The Socialist Invasion of Baja California, 1911." *Pacific Historical Review* 15, no. 3 (1946): 295–304. http://www.jstor.org/stable/3635477.
Gómez Estrada, José Alfredo. *Gobierno y casinos: El origin de la riqueza de Abelardo L. Rodríguez*. Mexicali, Baja California: Universidad Autónoma de Baja California, 2002.
González, Gilbert G. "Company Unions, the Mexican Consulate, and the Imperial Valley Agricultural Strikes, 1928–1934." *Western Historical Quarterly* 27, no. 1 (1996): 53–73.
———. *Culture of Empire: American Writers, Mexico, and Mexican Immigrants, 1880–1930*. Austin: University of Texas Press, 2004.
———. *Mexican Consuls and Labor Organizing: Imperial Politics in the American Southwest*. Austin: University of Texas Press, 1999.
Gray, James. *The American Civil Liberties Union of Southern California and Imperial Valley Agricultural Labor Disturbances, 1930, 1934*. San Francisco: R. and E. Associates, 1977.
Gregory, James N. *American Exodus: The Dust Bowl Migration and Okie Culture in California*. New York: Oxford University Press, 1989.

Guerin-Gonzales, Camille. "The International Migration of Workers and Segmented Labor: Mexican Immigrant Workers in California Industrial Agriculture, 1900–1940." In *The Politics of Immigrant Workers: Labor Activism and Migration in the World Economy since 1830*, edited by Camille Guerin-Gonzales and Carl Strikwerda, 155–74. New York: Holmes and Meier, 1993.

———. *Mexican Workers and American Dreams: Immigration, Repatriation, and California Farm Labor, 1900–1939*. New Brunswick, NJ: Rutgers University Press, 1994.

Gutiérrez, David G. "Migration, Emergent Ethnicity, and the 'Third Space': The Shifting Politics of Nationalism in Greater Mexico." *Journal of American History* 86, no. 2 (1999): 481–517.

Gutiérrez, Ramón A. *When Jesus Came, the Corn Mothers Went Away: Marriage, Sexuality, and Power in New Mexico, 1500–1846*. Stanford, CA: Stanford University Press, 1991.

Haas, Lisbeth. *Conquests and Historical Identities in California, 1769–1936*. Berkeley: University of California Press, 1996.

Hall, Greg. *Harvest Wobblies: The Industrial Workers of the World and Agricultural Laborers in the American West, 1905–1930*. Corvallis: Oregon State University Press, 2001.

Hall, Linda B. "Alvaro Obregón and the Politics of Mexican Land Reform, 1920–1924." *Hispanic American Historical Review* 60, no. 2 (1980): 213–38. http://www.jstor.org/stable/2513216.

Hall, Linda B., and Don M. Coerver. *Revolution on the Border: The United States and Mexico, 1910–1920*. Albuquerque: University of New Mexico Press, 1990.

Harris, Charles H., III, and Louis R. Sadler. *The Secret War in El Paso: Mexican Revolutionary Intrigue, 1906–1920*. Albuquerque: University of New Mexico Press, 2009.

Hart, John M. *Empire and Revolution: The Americans in Mexico since the Civil War*. Berkeley: University of California Press, 2002.

———. "The Mexican Revolution, 1910–1920." In *The Oxford History of Mexico*, edited by Michael C. Meyer and William H. Beezley, 435–65. New York: Oxford University Press, 2000.

Healey, Dorothy Ray, and Maurice Isserman. *California Red: A Life in the American Communist Party*. Urbana: University of Illinois Press, 1993.

Henderson, George L. *California and the Fictions of Capital*. Oxford: Oxford University Press, 1999.

Henke, Christopher R. *Cultivating Science, Harvesting Power: Science and Industrial Agriculture in California*. Cambridge, MA: MIT Press, 2008.

Hernández, José A. *Mutual Aid for Survival: The Case of the Mexican American*. Malabar, FL: Robert E. Krieger, 1983.

Hernández, Kelly L. *Migra!: A History of the U.S. Border Patrol*. Berkeley: University of California Press, 2010.

Herrera Carrillo, Pablo. *Reconquista y colonización del Valle de Mexicali y otros escritos paralelos*. Mexicali, Baja California: Universidad Autónoma de Baja California, 2002.

Hosmer, Helen. "Imperial Valley: Triumph and Failure in the Colorado Desert." *American West* 3, no. 1 (1966): 34–49, 79.

Howe, Edgar F., and Wilbur J. Hall. *The Story of the First Decade in Imperial Valley, California*. Imperial, CA: Edgar F. Howe and Sons, 1910.

Hu-DeHart, Evelyn. "The Chinese of Baja California Norte, 1910–1934." *Proceedings of the Pacific Coast Council on Latin American Studies* 12 (1985–86): 9–30.

Hundley, Norris, Jr. "Clio Nods: Arizona *v.* California and the Boulder Canyon Act—A Reassessment." *Western Historical Quarterly* 3, no. 1 (1972): 17–51.

———. *Dividing the Waters: A Century of Controversy between the United States and Mexico*. Berkeley: University of California Press, 1966.

———. *The Great Thirst: Californians and Water: A History, 1770s–1990s*. Rev. ed. Berkeley: University of California Press, 2001.

———. "The Politics of Reclamation: California, the Federal Government, and the Origins of the Boulder Canyon Act—A Second Look." *California Historical Quarterly* 52, no. 4 (1973): 292–325.

———. "The Politics of Water and Geography: California and the Mexican-American Treaty of 1944." *Pacific Historical Review* 36, no. 2 (1967): 209–26.

———. *Water and the West: The Colorado River Compact and the Politics of Water in the American West*. Berkeley: University of California Press, 1975.

Ichioka, Yuji. *The Issei: The World of the First Generation Japanese Immigrants, 1885–1924*. New York: Free Press, 1988.
Isenberg, Andrew C. *Mining California: An Ecological History*. New York: Hill and Wang, 2005.
James, George W. *The Wonders of the Colorado Desert*. Boston: Little, Brown, 1906.
Jameson, Elizabeth. *All That Glitters: Class, Conflict, and Community in Cripple Creek*. Urbana: University of Illinois Press, 1998.
Kelly, William H. *Cocopa Ethnography*. Tucson: University of Arizona Press, 1977.
Kennan, George. *The Salton Sea: An Account of Harriman's Fight with the Colorado River*. New York: Macmillan, 1917.
Kennedy, David M. *Freedom from Fear: The American People in Depression and War, 1929–1945*. New York: Oxford University Press, 1999.
Kirchner, John A. *Baja California Railways*. Los Angeles: Dawson's Book Shop, 1988.
Knobloch, Frieda. *The Culture of Wilderness: Agriculture as Colonization in the American West*. Chapel Hill: University of North Carolina Press, 1996.
Koppes, Clayton R. "Public Water, Private Land: Origins of the Acreage Limitation Controversy, 1933–1953." *Pacific Historical Review* 47, no. 4 (1978): 607–36.
Kroeber, Clifton B. *Man, Land, and Water: Mexico's Farmlands Irrigation Policies: 1885–1911*. Berkeley: University of California Press, 1983.
Lavender, David. *California: Land of New Beginnings*. New York: Harper and Row, 1972.
Lee, Lawrence B. "William Ellsworth Smythe and the Irrigation Movement: A Reconsideration." *Pacific Historical Review* 41, no. 3 (1972): 289–311.
———. "William E. Smythe and San Diego, 1901–1908." *Journal of San Diego History* 19, no. 1 (1973): 10–24.
Leonard, Karen I. *Making Ethnic Choices: California's Punjabi Mexican Americans*. Philadelphia: Temple University Press, 1992.
Limerick, Patricia N. *Desert Passages: Encounters with the American Deserts*. Albuquerque: University of New Mexico Press, 1985.
———. *The Legacy of Conquest: The Unbroken Past of the American West*. New York: W. W. Norton, 1988.
Lines, William J. *Taming the Great South Land: A History of the Conquest of Nature in Australia*. Athens: University of Georgia Press, 1991.
López, Ian F. Haney. *White by Law: The Legal Construction of Race*. New York: New York University Press, 1996.
Loveridge-Sanbonmatsu, Joan. *Imperial Valley Nisei Women: Transcending Poston*. DeWitt, NY: New Persephone Press, 2006.
Lowitt, Richard. *The New Deal and the West*. Bloomington: Indiana University Press, 1984.
Majka, Linda C., and Theo J. Majka. *Farm Workers, Agribusiness, and the State*. Philadelphia: Temple University Press, 1982.
McWilliams, Carey. *California: The Great Exception*. Westport, CT: Greenwood Press, 1949.
———. *Factories in the Fields: The Story of Migratory Farm Labor in California*. 1935. Berkeley: University of California Press, 1999.
———. *Ill Fares the Land: Migrants and Migratory Labor in the United States*. 1941. New York: Barnes and Noble, 1967.
McWilliams, James E. *American Pests: The Losing War on Insects from Colonial Times to DDT*. New York: Columbia University Press, 2008.
Meeks, Eric V. *Border Citizens: The Making of Indians, Mexicans, and Anglos in Arizona*. Austin: University of Texas Press, 2007.
Mellinger, Phil. "How the IWW Lost Its Western Heartland: Western Labor History Revisited." *Western Historical Quarterly* 27, no. 3 (1996): 303–24.
Middlebrook, Kevin J. *The Paradox of Revolution: Labor, the State, and Authoritarianism in Mexico*. Baltimore: Johns Hopkins University Press, 1995.
Mitchell, Don. *The Lie of the Land: Migrant Workers and the California Landscape*. Minneapolis: University of Minnesota Press, 1996.

———. *They Saved the Crops: Labor, Landscape, and the Struggle over Industrial Farming in Bracero-Era California*. Athens: University of Georgia Press, 2012.

Mitchell, Greg. *The Campaign of the Century: Upton Sinclair's Race for Governor of California and the Birth of Media Politics*. New York: Random House, 1992.

Moeller, Beverley B. *Phil Swing and Boulder Dam*. Berkeley: University of California Press, 1971.

Molina, Natalia. "Constructing Mexicans as Deportable Immigrants: Race, Disease, and the Meaning of "Public Charge." *Identities: Global Studies in Culture and Power* 17 (2010): 641–66.

Moore, Stephen T. "Refugees from Volstead: Cross-Boundary Tourism in the Northwest during Prohibition." In *The Borderlands of the American and Canadian Wests: Essays on Regional History of the Forty-Ninth Parallel*, edited by Sterling Evans, 247–61. Lincoln: University of Nebraska Press, 2006.

Morrissey, Katherine G. *Mental Territories: Mapping the Inland Empire*. Ithaca, NY: Cornell University Press, 1997.

Nadeau, Remi. *The Water Seekers*. 1950. Santa Barbara, CA: Peregrine Smith, 1974.

Ngai, Mae M. *Impossible Subjects: Illegal Aliens and the Making of Modern America*. Princeton, NJ: Princeton University Press, 2004.

Olcott, Jocelyn. *Revolutionary Women in Postrevolutionary Mexico*. Durham, NC: Duke University Press, 2005.

Orsi, Richard J. *Sunset Limited: The Southern Pacific Railroad and the Development of the American West*. Berkeley: University of California Press, 2005.

Peck, Gunther. *Reinventing Free Labor: Padrones and Immigrant Workers in the North American West, 1880–1930*. Cambridge: Cambridge University Press, 2000.

Pisani, Donald J. *From the Family Farm to Agribusiness: The Irrigation Crusade in California and the West, 1850–1931*. Berkeley: University of California Press, 1984.

———. "The Irrigation District and the Federal Relationship: Neglected Aspects of Water History in the Twentieth Century." In *The Twentieth-Century West: Historical Interpretations*, edited by Gerald D. Nash and Richard W. Etulain, 257–92. Albuquerque: University of New Mexico Press, 1989.

———. *To Reclaim a Divided West: Water, Law, and Public Policy, 1848–1902*. Albuquerque: University of New Mexico Press, 1992.

———. *Water and American Government: The Reclamation Bureau, National Water Policy, and the West, 1902–1935*. Berkeley: University of California Press, 2002.

Preston, William, Jr. *Aliens and Dissenters: Federal Suppression of Radicals, 1903–1933*. New York: Harper Torchbooks, 1966.

Raat, W. Dirk. *Revoltosos: Mexico's Rebels in the United States, 1903–1923*. College Station: Texas A&M University Press, 1981.

Recio, Gabriela. "Drugs and Alcohol: US Prohibition and the Origins of the Drug Trade in Mexico, 1910–1930." *Journal of Latin American Studies* 34, no. 1 (2002): 21–42.

Reisler, Mark. "Always the Laborer, Never the Citizen: Anglo Perceptions of the Mexican Immigrant during the 1920s." *Pacific Historical Review* 45, no. 2 (1976): 231–54. www.jstor.org/stable/3638496.

———. *By the Sweat of Their Brow: Mexican Immigrant Labor in the United States, 1900–1940*. Westport, CT: Greenwood Press, 1976.

Robbins, William G. *Landscapes of Promise: The Oregon Story, 1800–1940*. Seattle: University of Washington Press, 1997.

Robinson, W. W. *Land in California*. Berkeley: University of California Press, 1948.

Rodríguez, Abelardo L. *Memoria administrativa del gobierno del Distrito Norte de la Baja California, 1924–1927*. 1928. Mexicali, Baja California: Universidad Autónoma de Baja California, 1993.

Roediger, David R. *The Wages of Whiteness: Race and the Making of the American Working Class*. Rev. ed. London: Verso, 1999.

Romero, Robert Chao. *The Chinese in Mexico, 1882–1940*. Tucson: University of Arizona Press, 2010.

Round, Philip H. *The Impossible Land: Story and Place in California's Imperial Valley*. Albuquerque: University of New Mexico Press, 2008.

Ruiz, Vicki L. *Cannery Women, Cannery Lives: Mexican Women, Unionization, and the California Food Processing Industry, 1930–1950*. Albuquerque: University of New Mexico Press, 1987.

———. *From Out of the Shadows: Mexican Women in Twentieth-Century America*. New York: Oxford University Press, 1998.

Russell, Edmund. *War and Nature: Fighting Humans and Insects with Chemicals from World War I to Silent Spring*. Cambridge: Cambridge University Press, 2001.

Sackman, Douglas C. *Orange Empire: California and the Fruits of Eden*. Berkeley: University of California Press, 2005.

Samaniego López, Marco Antonio. "El control del río Colorado como factor histórico la necesidad de estudiar la relación tierra/agua." *Frontera Norte* 20, no. 40 (2008): 49–78.

Sánchez, George J. *Becoming Mexican American: Ethnicity, Culture and Identity in Chicano Los Angeles, 1900–1940*. New York: Oxford University Press, 1993.

Sánchez Ramírez, Oscar. *Crónica agrícola del Valle de Mexicali*. Mexicali, Baja California: Universidad Autónoma de Baja California, 1990.

Sanderson, Steven E. *Agrarian Populism and the Mexican State: The Struggle for Land in Sonora*. Berkeley: University of California Press, 1981.

Sandos, James A. "Northern Separatism during the Mexican Revolution: An Inquiry into the Role of Drug Trafficking, 1910–1920." *The Americas* 41, no. 2 (1984): 191–214.

———. *Rebellion in the Borderlands: Anarchism and the Plan of San Diego, 1904–1923*. Norman: University of Oklahoma Press, 1992.

Sano, Iwao Peter. *One Thousand Days in Siberia: The Odyssey of a Japanese-American POW*. Lincoln: University of Nebraska Press, 1997.

Santiago, Myrna I. *The Ecology of Oil: Environment, Labor, and the Mexican Revolution, 1900–1938*. Cambridge: Cambridge University Press, 2006.

Saund, D. S. *Congressman from India*. New York: E. P. Dutton, 1960.

Schantz, Eric M. "All Night at the Owl: The Social and Political Relations of Mexicali's Red-Light District, 1913–1925." *Journal of the Southwest* 43, no. 4 (2001): 549–602. www.jstor.org/stable/40170171.

Schonfeld, Robert G. "The Early Development of California's Imperial Valley: Part I." *Southern California Quarterly* 50, no. 3 (1968): 279–307.

———. "The Early Development of California's Imperial Valley: Part II." *Southern California Quarterly* 50, no. 4 (1968): 395–426.

Shah, Nayan. *Stranger Intimacy: Contesting Race, Sexuality, and the Law in the North American West*. Berkeley: University of California Press, 2011.

Smith, Michael M. "Andrés G. García: Venustiano Carranza's Eyes, Ears, and Voice on the Border." *Mexican Studies/Estudios Mexicanos* 23, no. 2 (2007): 355–86. http://www.jstor.org/stable/10.1525/msem.2007.23.2.355.

Smythe, William E. *The Conquest of Arid America*. Rev. ed. New York: Macmillan, 1905.

Snodgrass, Michael. '"We Are All Mexicans Here': Workers, Patriotism, and Union Struggles in Monterrey." In *The Eagle and the Virgin: Nation and Cultural Revolution in Mexico, 1920–1940*, edited by Mary Kay Vaughan and Stephen E. Lewis, 314–34. Durham, NC: Duke University Press, 2006.

Sperry, Robert L. "When the Imperial Valley Fought for Its Life." *Journal of San Diego History* 21, no. 1 (1975): 1–25.

Spickard, Paul. *Almost All Aliens: Immigration, Race, and Colonialism in American History and Identity*. New York: Routledge, 2007.

St. John, Rachel. *Line in the Sand: A History of the Western U.S.-Mexico Border*. Princeton, NJ: Princeton University Press, 2011.

Starr, Kevin. *Endangered Dreams: The Great Depression in California*. New York: Oxford University Press, 1996.

Stein, Walter J. *California and the Dust Bowl Migration*. Westport, CT: Greenwood Press, 1973.

Stoll, Steven. *The Fruits of Natural Advantage: Making the Industrial Countryside in California*. Berkeley: University of California Press 1998.

Stout, Joseph A., Jr. *Schemers and Dreamers: Filibustering in Mexico, 1848–1921*. Fort Worth, TX: Texas Christian University Press, 2002.

Stowe, Noel J. "Pioneering Land Development in the Californias: An Interview with David Otto Brant." *California Historical Society Quarterly* 47, no. 1 (1968): 15–39.

Street, Richard S. *Beasts of the Field: A Narrative History of California Farmworkers, 1769–1913*. Stanford, CA: Stanford University Press, 2004.

Suarez-Potts, William J. *The Making of Law: The Supreme Court and Labor Legislation in Mexico, 1875–1931*. Stanford, CA: Stanford University Press, 2012.

Sykes, Godfrey. *The Colorado Delta*. Washington, DC: Carnegie Institution; New York: American Geographical Society, 1937.

Taylor, Lawrence D. "Fighting the Patria on the 'Wrong' Side of the Border: The Role of the Mexican Consuls in the Federal Campaign against the Magonista Rebels in Baja California in 1911." *Journal of Borderland Studies* 22, no. 2 (2007): 1–20.

Taylor, Paul S. *Mexican Labor in the United States: The Imperial Valley*. Berkeley: University of California Press, 1930; rpt. New York: Arno Press and the *New York Times*, 1970.

———. "Mexican Migration and the 160-Acre Water Limitation." *California Law Review* 63, no. 3 (1975): 732–50.

———. "Water, Land, and Environment [in] Imperial Valley: Law Caught in the Winds of Politics." *Natural Resources Journal* 13, no. 1 (1973): 1–35.

Teisch, Jessica B. *Engineering Nature: Water, Development, and the Global Spread of American Environmental Expertise*. Chapel Hill: University of North Carolina Press, 2011.

Tobler, Hans W. "Peasants and the Shaping of the Revolutionary State, 1910–40." In *Riot, Rebellion, and Revolution: Rural Social Conflict in Mexico*, edited by Friedrich Katz, 487–518. Princeton, NJ: Princeton University Press, 1988.

Truett, Samuel. *Fugitive Landscapes: The Forgotten History of the U.S.-Mexico Borderlands*. New Haven, CT: Yale University Press, 2006.

Tyrrell, Ian. *True Gardens of the Gods: California-Australian Environmental Reform, 1860–1930*. Berkeley: University of California Press, 1999.

Vanderwood, Paul J. *Disorder and Progress: Bandits, Police, and Mexican Development*. Revised and enlarged ed. Wilmington, DE: SR Books, 1992.

———. *Satan's Playground: Mobsters and Movie Stars at America's Greatest Gaming Resort*. Durham, NC: Duke University Press, 2010.

Vaught, David. *Cultivating California: Growers, Specialty Crops, and Labor, 1875–1920*. Baltimore: Johns Hopkins University Press, 1999.

Velázquez Morales, Catalina. *Los inmigrantes chinos en Baja California, 1920–1937*. Mexicali, Baja California: Universidad Autónoma de Baja California, 2001.

Vollmann, William T. *Imperial*. New York: Viking, 2009.

Walker, Richard A. *The Conquest of Bread: 150 Years of Agribusiness in California*. New York: New Press, 2004.

Walsh, Casey. *Building the Borderlands: A Transnational History of Irrigated Cotton along the Mexican-Texas Border*. College Station: Texas A&M University Press, 2008.

Ward, Evan R. *Border Oasis: Water and the Political Ecology of the Colorado River Delta, 1940–1975*. Tucson: University of Arizona Press, 2003.

———. "The Twentieth-Century Ghosts of William Walker: Conquest of Land and Water as Central Themes in the History of the Colorado River Delta." *Pacific Historical Review* 70, no. 3 (2001): 359–85.

Weber, Devra. *Dark Sweat, White Gold: California Farm Workers, Cotton, and the New Deal*. Berkeley: University of California Press, 1994.

———. "Keeping Community, Challenging Boundaries: Indigenous Migrants, Internationalist Workers, and Mexican Revolutionaries, 1900–1920." In *Mexico and Mexicans in the Making of the United States*, edited by John Tutino, 208–35. Austin: University of Texas Press, 2012.

Whitten, Woodrow C. *Criminal Syndicalism and the Law in California: 1919–1927*. Philadelphia: American Philosophical Society, 1969.

Williams, Anita Alvarez. *The Cocopah People*. Phoenix: Indian Tribal Series, 1974.

———. "People and the River." *Journal of the Southwest* 39, nos. 3/4 (1997): 331–51.

Woeste, Victoria S. *The Farmer's Benevolent Trust: Law and Agricultural Cooperation in Industrial America, 1865–1945*. Chapel Hill: University of North Carolina Press, 1998.

Wollenberg, Charles. "*Huelga*, 1928 Style: The Imperial Valley Cantaloupe Workers' Strike." *Pacific Historical Quarterly* 38, no. 1 (1969): 45–58.

Woodbury, David O. *The Colorado Conquest*. New York: Dodd, Mead, 1941.

Worster, Donald. *Rivers of Empire: Water, Aridity, and the Growth of the American West*. New York: Oxford University Press, 1985.

Wright, Harold Bell. *The Winning of Barbara Worth*. Chicago: Book Supply, 1911.

Wrobel, David M. *Promised Lands: Promotion, Memory, and the Creation of the American West*. Lawrence: University Press of Kansas, 2002.

Wyckoff, William. *Creating Colorado: The Making of a Western American Landscape, 1860–1940*. New Haven, CT: Yale University Press, 1999.

Wyman, Mark. *Hoboes: Bindlestiffs, Fruit Tramps, and the Harvesting of the West*. New York: Hill and Wang, 2010.

Index

50 percent (later 80 percent) jobs for Mexicans, Mexicali, 74, 99, 125, 160
160-acre limitation: debate within Department of Interior, 36; Harry Chandler opposed it, 33; Imperial Irrigation District legal opinion, exemption, 36, 38; Department of Interior questioned legality of IID's exemption, 37; lawsuit seeking enforcement of, 165; non-compliance of, 165; political turmoil over, 36; resistance to, 29

absentee landowners, 45–46, 66, 68, 167
acala, cotton, 92
acreage, Imperial Valley, contemporary, 166; limitation, federal irrigation project, 166
Adams, Frank: lower Colorado River, water rights report, 34
Adams, R. L., 112
adolescence obesity and diabetics, 164
agitators, 135–136
agrarian, Mexico: department, 154–155; law, 131–132; militants, 161; resettlement, 140
agraristas: land petitions, 153; land redistribution, 74
agribusiness, elite: local authority, 47; imperialism, 174
Agricultural Adjustment Administration (AAA), 95
Agricultural Workers' Industrial League (AWIL), 130, 139–142
Agricultural Workers' Organization, 130. *See also* Industrial Workers of the World
Agriculture: 1910s, 73–75; and livestock production, contemporary, Imperial County, 165, 167; census statistics, 75–76; commissioner reports, contemporary, 168; crop associations, 73; double cropping, 76; experiment station, 91; farm managers, 70; first decade, 69–70; international reputation, 76; labor-intensive, 68; irrigation, 9; overproduction, 92; racial, 68; unsustainable, 163
Agricultural Conservation Association, 95–96
Air Pollution Control District, 171
air quality, 170–171, 174; asthma and respiratory ailments, 164; violations, 164
Alamo Canal, 11, 16; floods 1905–1907, 19; magonista threat, 27; purchased by Imperial Irrigation District, 30
Alamo River: drainage canyon, 43; pollution, 164
Algodones: magonista invasion, 27; troops rebelled, 131
Alien Land Law (1913), 69, 71, 77
Alien Land Law (1920), 51, 69, 71–72, 83–86: African Americans recruited, 86; Asians relocated after, 85; evasion of, 84–86; impact of, 85, legacy of, 87–88; support of, 84–85; testing legality of, 86
alien land laws, 79; evasion of, 78; legacy, 96
alkali: drainage, 60–61; excess salt, 59
All-American Canal, 34, 38, 93, 96, 122, 159–160, 165–166: amount of water, 60; court validation, 36; critics, 96; federal bonds, 70; funding, 120; funding celebrations, 36–37; jobs, 121; opposition, 12; proposal, 29; river control, 11–12; federal ownership, 8; wrestle river from Mexico, 33
Allison, Chester, Mexicali Valley, Japanese colonization of, 87
Almazán, Juan Andrew, 153
American-Mexican Claims Commission, 155
American Civil Liberties Union (ACLU): 10, 128, 157–158; encouraged fieldworker union, 150; Los Angeles, 141; publicized violations of civil liberties, 146; testimony of civil liberties violations, 151
American Federation of Labor (AFL): 100; agricultural union, 149–150; support of Alien Land Law (1920), 84; testimony to investigators, 151
American Fruit Growers, 85
American Legion, 51, 92, 116, 141, 149; charity, 119; support of Alien Land Law (1920), 86; protested use of school auditorium, 84
American Lung Association, 164
Andrade, Alberto, 20, 23, 26,
Andrade, Guillermo, 13, 43

annexation, rumors, 29, 153; resolution, Arizona senator, 153
Anti-Communism Association, 149–150, 158
anti-picketing ordinance, 147–148
aqueduct, 6. *See also* Alamo Canal
Aragón, Edmundo, 139–141
arbitration boards, 133, 157
Ariza, Carlos, 134
Arizona senator, 3
Arizona v. California (1963): 165–166; supplemental decree, 166, 174
Article 27, Mexico Constitution, ownership of natural resources, 30
Asians: social restraint, 77–78; tenants replaced, 51
Associated Farmers of California, 10, 92–93, 122, 128, 149, 158, 161
Associated Labor Bureau, 110–111, 138
Associated Press, lackluster economy, 163
Azteca Hall: 143–144, American Civil Liberties Union speakers, 146, 150; raid, 145

Bailey, Agnes, union spokesperson, 151
Baja California, 16; magonistas, 27, 129; annexation rumors, 29; Japanese colonization of, 87; social stability, 164; territorial status, 153
Balfour-Guthrie, 50
Bank of America, 46
Bank of Italy, 46
Barkley, Winfield, 3
beef plant closing, Brawley, 167–168
Beltrán y Puga, Fernando, 25
birds, migratory, 170
blackbirds, 64
Bloch, Louis: Mexicans refused charity, 119; strike investigation, 111, 113, 137–138
Board of Supervisors, Imperial County: anti-Japanese ordinance, 167; appointed Imperial Irrigation District board, 46–47; complaint to governor, 144
bolshevikis, 131
bonus (holdback), 113, 138, 156
boosters, 42; colonization, 4, 12–13; desert Garden of Eden, 11; sanitary housing, 114
border closing, 9 pm, 55
Border Patrol, 110, 135; creation, 69
Boulder Canyon Project Act, 34, 38–39, 166; court validation, 36
Boulder Dam (Hoover Dam), 11, 34, 153; importance, 9
Bureau of Agricultural Economics, pooling and clearinghouse, 92
Bureau of Immigration, 102, 105
Bureau of Land Management, 175
Bureau of Markets, 93–94
Bureau of Reclamation, 18, 36, 38; Boulder Dam, 34; river control, 39; water rights report, Frank Adams, 34; water shortage, San Diego conference, 174; watershed report, 173–174
Bureau of Standardization, 94
Braceros, transnational migrants, 99
Bracero Program, 168–169
Brawley, 20, 52, 58, 84; Industrial Workers of the World, 129–130; most polluted town in California, 164; rabbit hunt, 65; segregation, 56; high school farm trainee program, 51
Brawley News, 45, 84–85, 87, 94, 98, 109, 120, 124, 145–146
British imperial system, 76
Bromley, David, 159

Cable Act, 85
Calexico, 52, 69, 132–133; All-American Canal route, 37; creation, 16; Industrial Workers of the World, 129; port of entry, 52; flood, 22; troops stationed, 27, 29, 131
Calexico Chronicle, 132; Colorado River Land Company proxy, 33; gopher control, 65; police chief, control, 56
California, 68; wetlands, 170; California Citizens' Committee, 123–124
California Cotton Exchange, importing Mexican labor, 91
California Development Company, 11, 14, 66, 166; agricultural experiment station, 60; acquiring land, 45; cooperation with Southern Pacific Railroad, colonization, 42; created Imperial Land Company, 41; creation, 13; labor, 99; 1904 Mexican water conveyance concession, 38; mutual water companies, 44; opposed Reclamation Act, 18; receivers hated each other, 28–29; receivership, 38; transnational water company, 20; white farmers, 41
California legislature: approved All-American Canal contract, 35; labor regulations, 9–10; regulated agriculture, 160
California-Baja California borderland, 161
California-Mexico Land and Cattle Company (C&M ranch), Calexico, 60; creation, 43
Calipatria, 120–121; Industrial World of the World, 130
Calles, Plutarco Elías: immigration, 74; private land ownership, 74
Cameron, J., 90
campesinos, Mexicali, 74
Canadian Pacific Coast, 76

canals, Mexicali Valley, 74
Cannery and Agricultural Workers Industrial Union (CAWIU), 142–152, 158; collective bargaining request, 144; strike defeat, 152. *See also* strikes
cantaloupes (melons), 75; Japanese tenants, 75–76; statistics, 75
Cantú, Esteban, governor (1915–1920): Asian immigration, 103; bribery, 52–53; cavalry and militia, 131; agricultural policies, 52; licensed vice, 52–53; Mexican labor recruitment, 105; military buildup, 53, modernized Mexicali, 53; promoted Chinese tenancy and labor, 52; road to Tijuana, 102; stranding Chinese transmigrants, 103–104
Cárdenas, Lázaro, 161; Colorado River Land Co. contract, 154; Colorado River Land Co. expropriation, 158, 172–173; land expropriation, 116; met with ejidatarios in Mexico City, 154; radio address, 154; Mexican repatriation, 116, 153; aid telegrams to 153; visited Mexicali, 153
census (1910), 51–52; farm acreage and racial farmers, 70
census (1920), 51–52
Central Labor Council, of American Federation of Labor, 144, 149
Chaffey, George, 45, 48, 114–117; water stock, 43
chain gang, 55, 57, 132
Chamber of Commerce, Imperial County, 110
Chamber of Commerce, Mexicali, labor bureau, 110
Chamber of Commerce, United States, *Nation's Business*, 89
Chandler, E. E., absentee landowner, 71–72; racial opinions, 77
Chandler, Harry, 43, 50, 160; anti-labor and anti-union, 131. *See also* California-Mexico Land and Cattle Company. *See also* Colorado River Land Company; cotton cooperative, 91; criticized by Hiram Johnson, 33. *See also* Imperial Valley Farm Lands Association; labor recruitment, 104; lobbying against Mexico quota, 116
Chaplin brothers, 49
charity, 118; Brawley, 119–120; Mexicans, 110
Chase, J. Smeaton, 2
Chávez, César, 169. *See also* United Farm Workers
chemicals. *See* poisons
Chikahisa, Kayo, 78
Chikahisa, Ray, 78
Chinese, tenants and pickers, Mexicali Valley, 43, 68, 73–74; repatriation, 154; credit, 72; migrants, 9
Ciudad Juárez, 27
civil liberties violations, 141. *See also* American Civil Liberties Union
Civil Works Administration, 121
Clark, H. H., 160; cotton cooperative, 91; Mexican labor recruitment, 105; president Calexico Chamber of Commerce, 91–92; president California Cotton Growers Association, 92
climate, 4; change, 163
Club Demócrata, 143
College of Agriculture, University of California, Berkeley, research station, Calexico, 60
collective bargaining, 134, 143–144; agribusiness refusal of, 139–140, 151; shed union, 156
colonization, 1, 6, 9; financial difficulties, 13–14; Mexicali Valley, 154
colonizing, desert, 58–61, 66; feminine power, 40
Colorado Desert, 4, 12, 16; contradictions, 1–2; feminine images, 1; negative connotations, 1
Colorado River, author recollection, 176; battle over, 8; economic benefit of, 175; changed course, 22, 31; contest for, 161; drought and potable water, 37; federal control, 8; floods (1905–1907), 19–25; Imperial Irrigation District allocation, 169; Interstate Compact, 32; sediment, 4; states met to apportion, 31–32; subjugation, 3–4, 38; US opposed to Mexico over, 160; US-Mexico water treaty (1944), 37–38, 164; US intended to control, 153–154; water rights, 8
Colorado River Commission, 32
Colorado River delta, 5–6, 172–173, 175
Colorado River Land Company, S.A., 91, 125; agricultural advisers, 60; anti-Mexican farmer and worker policies, 43, 72–73; Asian tenancy, 43; avoided land confiscation, 74; avoided land sales to Mexicans, 74; from cattle ranching to cotton farming, 43, 73; Chinese tenancy and labor, 52; contract to sell plots, 154; cooperated with Imperial Valley farmers, 68; cooperative, 91; cotton kingdom, 73; evicting braceros, 104; expropriation, 154–158; farm and labor policies, 52; complying with 50 percent (later 80 percent law), 74; complained high taxes, 52; hosted Mexican water commission, 25; importing Chinese, 73; Japanese colonization, 87; levee, 26; Mexican labor, 74; Mexican labor recruitment, 105; Mexican opposition of, 131; Mexico's proxy against Boulder Canyon Project Act, 33; opposed Imperial Irrigation District's anti-Mexican water user policies, 33; overview, 9; owners and creation, 43; permit to draw water from river, 30–31; releasing Chinese migrations, 103; subject of political

Colorado River Land Company, S.A. (*cont.*) attacks, 33; tenants, 72; tenants, Asian, 68; tenants, credit, 72. *See also* Chinese
Comité Mexicano de Bienestar report, 147
commerce, contemporary, 165
Commission of Immigration and Housing, 122, 124
Commonwealth Club, 33
communists, 10, 141, 147; labor organizers, 128, 139–142
Compañia de Terrenos y Aguas de la Baja California, S.A., 26; creation, 38; dissention, receivers, 28–29; Imperial Irrigation District in Mexico, 33; owned by Imperial Irrigation District, 30; regulated by Mexico, 32; water rights, 34
Confederación de Uniones Obreros Mexicanos (CUOM), 133
Congress, US, approved $1 million for levees, 26
Conkling, Marvin: Alien Land Law (1920), 86; lawyer, 46
Connell, Charles, US Conciliation Service, 140
conquerors, 1
conquest, Colorado River, 25; of landscape, 4; ideology, 19, 40, 159–160
Conquest of Arid America, 14
consul, Chinese, 154
consul, Japanese, 86
consul, Mexico, 25, 128, 134, 138–141, 143; *bracero* union, 148–153; *bracero* union contract, 150; Brawley high school auditorium, 149. *See also* Joaquín Terrazas; collective bargaining, 150, 152–153; Mexican government's instructions, 138, 145; *mutualistas*, 119; San Francisco, 13
consul, US, 143
contract labor system, 157
control, 129; by agribusiness elite, 47; Colorado River, 1, 8, 28, 39; ideological roots, 28; ideology, ix, 39–41; ideology, legacy, 174; mindset of colonizers, 53–54; of dissenters, 6; of labor, 98, 115–118, 122, 124–126; of overproduction, 93–94; of nature, 1, 67; of nature for irrigated agriculture, 6; of nature, workers, and society, 4; of packing shed unionists, 155–157; of white migrants, 168. *See also* vagrancy; with labor saturation, 128; with pest eradication, 61–66; with poison, 62–63; with quarantine, 62; with racial segregation, 9; with seasonal labor program, 104–106; with vagrancy law, 56
convict labor, 57
Cooperative Movement among Japanese Farmers in California (1925), 91
cooperatives, 90–92
Cory, Harry, 22–24
corruption, police and public officials, 54
cotton, 75; agriculturalists, Arizona, 105; movie, plantation South, 73
Cotton Growers Association, California, 92
Cotton Growers Association, Imperial County, 104–106, 125
cotton industry, Mexicali, 74–75
Creator (God), 2
credit, 70–73, Los Angeles banks, 70
criminal syndicalism, 133, 141, 157–158; laws, 130
crop and livestock associations, 68, 90–93, 97
crop duster (airplane), 62
crop dusting, 8
crops: monthly harvest, 106–107; rotation, 76
Cruz Azul Mexicana, 119
Cucapá, 6, 16–17, 48, 175; contemporary issues, 172–173; displaced by floods, 25; floated to Salton Sea, 22; guide, 14; magonista invasion of Mexicali, 27

Delta Investment, 15–16
Democrats, 143
Department of Agriculture: blackbird control, 64; pest control, 62; soil survey, 17; subsidies, 93, 96; technical assistance, 60; Wildlife Services, pest eradication, 168
Department of Commerce, report (1923), 72–73
Department of Horticulture, origin, 91
Department of Industrial Relations, 114–115, 137
Department of Interior, 8, 35, 50, 96; All-American Canal contracts, 35, 37; 160-acre noncompliance, Imperial Irrigation District, 165; 160-acre limitation lawsuit, 165; opposed California Development Company, 17; Reclamation Service, 18; refused to confirm land patents, 25–26; Salton Sea sump, 60
Department of Labor, Conciliation Service, 140
Department of State, opposed Japanese colonization, Mexicali, 34
Department of Treasury, declined to sell bonds to build All-American Canal, 31
Department of War, authority over Colorado River per treaty with Mexico, 13, 18
deportation, during strikes, 135, 138; to Mexicali, homelessness, 165
desalination plant and drainage canal, Yuma, Arizona, 172
desert: agriculture, 68; Eden, 42; transformed into farms, 40
Desert Land Act, 42, 51; violations, 18, 42
desilting plant, on All-American Canal, 61
Díaz, Porfirio: exile, 27; irrigation studies, 25; magonista seizure of Mexicali, 27; persecuted critics, 27

discrimination, 51
disturbing the peace, 141; ordinance, 129
distrust of outsiders and regulations, 6
Duryea, Edwin, 22
dust storms, 171

Earl Fruit, 71, 85
east mesa, All-American Canal route, 29
Eden, desert, 14. *See also* Garden of Eden
egalitarian myth of colonization, 47
ejidatarios, 158; agricultural expansion, 164; land invasion, 154
ejidos, 153, 155, 158
El Asalto a las Tierras, 154
El Centro, 46, 52, 94, 120, 140; county seat, 46; economic stress, contemporary, 163–164; Industrial Workers of the World, 130; mutualista, 133; ozone unhealthfulness, 164; caravan to strike rally, 144; segregation, 56
El Hotel del Migrante Deportado, 165
El Mercurio, 34
El Nuevo Mundo, 136
El Universal, 136
Elks Lodge, Brawley, 1
Emigrant Trail, 12
employment office, 94, 112
Ensenada, 27, 52
environment, as commodity, 3; transformation for agriculture, 1, 3
Environmental Working Groups' Farm Subsidy Database, 168
Espionage and Selective Service Acts, 130
Estrada, Juan, 109, 137

Factories in the Fields, 124
Fair Labor Standards Act, 156
fallowing, 95, 97
farm: acreage, 70; experiment station, 94–95; factories, 87–90; landowners cooperation, Imperial-Mexicali Valley, 8
Farm Bureau, 51; Alien Land Law (1920), 84; anti-Asian, 79; hostility toward South Asians, 77; origin, 91; technical assistance, 60
farmers, 79, 81–82; African American (black and Negro), 78, 82, 86; Asian, census statistics, 77; Asian success, 77; Filipinos, 79, 82; farmers of color, 72–73, 76–82; capital, 70; contacts, 72, contracts, 71–72; credit, 70–71; institutes, 91; managers, census statistics, 88; contemporary, Imperial Valley, 166; opposed Japanese colonization of Mexicali Valley, 34; difficulty, 70, 96; five "C"s, 70–73
fascism, 152; letter to *Nation* about, 157; mentality, 161

Fawcett, W. S., 108
Federal Emergency Relief Administration, 120
Federal Employment Bureau, 147
federal farm bank, 46, 70
Federal Transient Service, 120, 122
Federation of Mexican Societies of Southern California, 133
fence, international border, 165
Fiege, Mark, rabbit control, Idaho, 65–66
Filipinos, 101–102; farmers, 79, 82; housing conditions, 115; jobs, 122; repatriation, 69
First Thirty Years, 45
floating sentence, 137
Flores Magón, Enrique and Ricardo, 27. *See also* magonistas
Florida fruit fly, 62
Fresno, CA, 100, 111, 129
fresno, land lever, 58
Fruit and Vegetable Workers Union (FVWU), strike, 155–156
Fuller, H., 90

García Peña, Ángel, 22
Garden of Eden, 2–4, 11, 50
Gerrard, S. A., 89; altering landscape, 58
Giannini Foundation of Agricultural Economics, 112
Gillette, Charles, 135–136, 139
Giraudo, Marie Antoinette, 78–79
Glassford, Pelham, 155; enemy of agribusiness, 152; American Civil Liberties Union opposition to, 152; anti-communism, 151; biography, 150–152; conciliation, 151; Mexican consul's union a farce, 152; opposed to Cannery and Agricultural Workers Industrial Union (CAWIU), 151; press releases, 152
Glen Canyon Dam, 172
González, Filemon, 134–136
González, Gilbert, 108
Goar, L. Gordon, 94–95
gopher eradication, 64–65
Grapes of Wrath, 124
Great Depression, 35; crime, 117; farmers of color, 81–82; Great Salt Lake, 22
Gregory, James, 118
Griffin, H. B.: leased farm to Japanese, 145; Mexicans plundered his farm, 57; $1 a day adequate for Mexican laborers, 146
growers-shippers, 88–89
Gulf of California, 3–5, 12
gunmen, during strike, 141

H-2A visas, 169
Half-Century Association, 103

Hall, Wilbur, 40
Hamaker, Eldrid, 156
Harvey, Harry, 139
Harriman, Edward (Southern Pacific Railroad), 21–22, 26; colonization assistance, 42; negotiated with Theodore Roosevelt, 24
Harris, J. A., 71
Harrigan, B. A.: absentee owner, 145; Horticultural Commissioner, 62–63, 65–66; starvation was good for migrants, 146
Harvey, Harry, 139
Hay Growers Association, 90
Heald, Elmer, 135–138, 141, 157–158; white farmer superiority, 96
health data, contemporary, 164
Heber, Anthony, 17; loan request, 21; Mexican water concession (1904), 19–20; opposed Reclamation Service, 19; rejected offer to buy California Development Company, 20–21
Heffernan, William, 13
Hidalgo Hall, Brawley, Mexican fieldworker union headquarters, 135
Highway Patrol, 144, 146
Hindu menace, 77
historical contribution of this study, 6, 8
historical revisionism, 49
hoboes, 115, 117
Holabird, W., California Development Company receiver, 26; receiver problems, 28–29
Holt, Leroy: banker, 46–47; treasurer, mutual water companies, 46
Holt, William: negotiated purchase California Development Company, 20; banker and businessman, 46–47
Holtville, 27, 52; creation, 46
Homestead Act, 42, 51, 71
homesteaders, 6
homestead laws, violations, 17
homesteading, 50
Hoover Dam, 38–39, 93, 96, 166. *See also* Boulder Dam
Hoover, Herbert, 36, 159
Horiuchi, Tsuji, 139
Horticultural Commissioner: leach alkali, 60. *See also* B. A. Harrigan
housing, labor, rural, 118, 125; improvements, 115; inspections, 114–115; unsanitary, 114–115 housing, labor, urban, unsanitary, 113–114
Howe, Edgar, 40

Ibarra, Modesto, 58–59
Ichikawa, Yoshio, 91
illiteracy, contemporary, 164
immigrants: likely to be a public charge, 119; self-help mutualistas, 119
immigration: Asian exclusion, US, 69; Chinese, Mexicali, 43, 100; circular migration, 140; contemporary, 165; opposition to quotas, Mexico, 138–139; undocumented Mexicans, 110; visa and head tax, 110; western hemisphere exemption, 110; local policy, 111
Immigration and Housing Commission, 99, 118
immigration restrictions, US, 102, 110
Immigration Service, 110
Imperial (town), 46, 52
Imperial Canal (Alamo Canal), 30
Imperial County: contemporary economy, 174–176; contemporary statistics, 163; creation, 48–49; reversal of reputation, 163
Imperial Dairyman's Association, 90–91
Imperial Dam, 34, 37
Imperial Irrigation District (IID), 12; aqueduct survey, 29; arguments for All-American Canal, 32; creation, 28, 46; critics of All-American Canal, 35; default on bonds, 35; depriving Mexico of water, 96; detrimental policies, 96; drainage, 59–60; farm titles during Depression, 36; favored US customers, 32; land concentration, 89; modernize distribution system, 175; negotiation to buy California Development Company, 29–30; 160-acre exemption, 37; 160-acre non-compliance, 165; overuse of water, 59; political turmoil, 31; purchased mutual water companies, 32; pro-All-American Canal circular, 35; public decisions, 28; public ownership, 38; requested Treasury bonds to build aqueduct, 31; solving problems, 31; state constitutional amendment own stock and property in Mexico, 29. *See also* Supreme Court; tracked water sales in Mexico, 59; water rights, Colorado River apportionment, 35; water rights, dubious, 165–166; none to Colorado River, 34
Imperial Land Company, 66; colonization, 15, 41–42; land grant fraud, 49; land grant violations, 18; water stock, 41–42, 44–45; traded water stock for labor, 44
Imperial Press, 11
Imperial Valley, named, 16
Imperial Valley Agricultural Association, Japanese, 84, 91, 132
Imperial Valley Farm Lands Association, 50
Imperial Valley Water Users Association, 19, 46
Imperial-Mexicali, twin towns at sea level, 4
Imperial-Mexicali Valley, ix, 4; description, 4–6
imperialistic ideology and language, 4
India, religious groups, 76

Indians, 6, 16; labor, 23–24
industrial agriculture, 90
internment, Japanese, 167
international border, 4; mapping, 12, 49
interracial alliances, 147
Interstate Commerce Law, 114
Interstate Compact, 32
irrigated agriculture, 68
Industrial Workers of the World (IWW), 10, 128–133, 157; legacy of repression, 28; magonistas, 27–28; membership, 130; philosophy, 129
Islas Marías, 140

Jack, Alvin: farm leader, 143; supported Alien Land Law (1920), 86; red baiting, 150
James, George Wharton, 1; floated to Salton Sea, 22
Japan, purchasing cotton, 75
Japanese: Alien Land Law (1920), 84; cooperatives, 91; farmers, 76–80, 82, 85; contemporary farming, 167; Imperial Valley Agricultural Association, 84, 91, 132; labor, 100; murder and repression, World War II, 167; packing shed owners, 91; repatriation, Mexicali, 154; tenants, Imperial Valley, 75–76; Mexicali Valley, 73; colonies, Mexicali, 33–34
Japanese-Mexican Labor Association, 100
Jefferson, Thomas, yeoman ideal, 3, 14, 40–41, 47, 54
Johnson, Ella Ruth, 6
Johnson, Hiram, 11; All-American Canal legislation, 32–33; critic of Harry Chandler and Colorado River Land Company, 33; 160-acre exemption, 36; opposed Japanese colonization in Mexicali Valley, 87

Kelley, Kevin, 171
Kennedy, David, 97
Kinoshita, S., 76
Knight, Paul, 156
Krantz, Tim, 171

labor, ethnic and racial groups: African Americans (blacks), 101; European immigrants, 100; Filipinos, 107, 101–102, 153; Indians, 99–100; Japanese, 100–101. *See also* labor; Mexicans, 96, 106; Mexicans and Filipinos, males, 106; multiracial, 41, 100; South Asians, 102; white and black, 115–118, 125, 153
labor, field, Imperial Valley: adult, male, migrants of color, 110; African Americans (blacks) moving to town, 101; agriculture exempt, reforms, 123; from Arizona, 104; children, 160; children, enforcing school attendance, 109–110; report of, 107; coalitions, 6; competition, 102; criticism of, 109; eliminated white migrants, 168; employer relations, 128–129; empowerment, 160–161; ethnic and racial competition, 109; exploitation and unpaid wages, 112–113; female, strike leadership, 144; during Great Depression, 115–118; grape girls, 111; ideal workforce by race, 124–125; improving wages and working conditions, 127; interracial union, 128; interracial strikes, 161; interracial unity, 139–142, 156; Imperial-Mexicali Valley housing, 160; lacked official power, 160; living conditions, 99, 112–115; Mexican circular transmigration, 104, 106; Mexican families, 109; Mexican importation, 91; Mexicans majority California agriculture, 106; Mexican migration, 102; migrants, contemporary, 175–176; New Deal, 143; petition waive immigration requirements for, 139; racial attitudes, 107–109; racial discord among, 108–109, 112, 116–117, 125, 139, 142, 156, 158; racial hierarchy, 6, 8–9; recruitment, 100, 102–106, 117; scarcity, 101–102; school attendance, 125; seasonal, 105–107; racially segmented, 98–99, 106–107; statistics, contemporary, 169; stereotypes, Mexicans and whites, 102–103. *See also* strikes; surplus, 104–106, 110–111, 139, 168; undocumented Mexicans, 168–169; union recognition, 144; wage reform, 112; white transients, 100; white supremacy, 116; working conditions, contemporary, 169
labor, field, Mexicali Valley, 68 Algodones anti-Asian mutiny, 103–104; anti-Asian, 103; Asians, 106; Chinese, 99, 125; Chinese exclusion, 125; Chinese, recruitment, 100; Colorado River Land Company, 100; 50 percent (later 80 percent) Mexicans, 125; official union, 140; rights, 160; statistics, 106; steamship transportation, 104, 106
labor, shed, 99, 106, 111–112, 127; anti-communism, 149; anti-field union, 138; anti-Mexican, 111–112; collective bargaining, 155–156; disempowered, 129; interracial discord, 142; Japanese, 111–112; non-strike bonus (holdback), 113, 138, 156; supported Alien Land Law (1920), 111; union organizers murdered, 157; unions, 27, 127; whites, 111–112
Labor Board of Baja California, Mexicali, 148–149
Labor Commissioner (CA), 94, 112–113, 144
labor contractors, 98, 102, 125, 128, 133, 157; abolition, 144; abuses, 113; contemporary, 169; regulation, 138; system, 115
labor force, four phases, 98–99

labor market, construction of, 98
Land Bureau: fraud of, 49; homesteaders, 42; Los Angeles office of, 44; policies, 12
land consolidation, 88–90
land redistribution, Mexicali Valley, 154
land script, 44–45
land seizure, Mexicali Valley, 131–132
large landowners, 66
large-scale farms, 9, 68
Last Frontier, 159
Laughrin, Richard: fluid race relations, 55; shaping landscape, 59
Lavender, David, 45
Leonard Commission, report, 146–147, 150
levees, 12, 25; unequal costs, 31
liberals, complaints of civil liberties violations, 145
Liera, Arnulfo, 104, 106
López, William, 137
Los Angeles Times, 33, 43, 139, 144
lunch program denied, 124

magonistas, 27–28
manaña culture, 108
maquiladoras, 164
martial law, Mexicali, 132
Martin, Mrs. Al, 50
McCullough, Campbell, 145–146
McWilliams, Carey, 97, 124, 161
Mejía, A., 140
Merriam, Frank, 121–122
Mexicali: agricultural boom (1950s), 172; arbitration board, labor, 152; avoided excessive bloodshed, contemporary, 165; census (1921), 53; creation, 16; magonista invasion, 27; official union, 140; port of entry, 52; sit-down strike, 154–155; troops stationed, 29; population growth, 164; Chamber of Commerce, 91
Mexicali Valley, 34, 91; annexation rumors, 29; American landowners, 68; Chinese labor, 68; economic diversity, 164–165; farm colonies, 153; federal oversight of Alamo Canal, 31; Mexican labor, 91; strike (1922), 132; troops crush arrest squatters, 154; post-World War II history, 164–165
Mexican farmers, 81–82; stereotypes, 72–73, 79, 81
Mexican Revolution, 102, 129, 133; militarized borderland, 131; seizure of Mexicali, 12
Mexicanizing Mexicali, 153–155, 158
Mexicans: credit, 72–73; demand land in Mexicali, 128–129; racially white, 51
Mexico, 58; agricultural department, 74; anti-communism, 134; 149; anti-strike policy, 135; arbitration boards, 133, 157; Colorado River Land Company expropriation, 154–155; consul's instructions, 138, 145; Constitution of 1917 (Article 27), 30, 131–132, 160; control expatriates, 133, 158; functionaries, 157; deprived of water by Imperial Irrigation District, 96; jousted with Imperial Irrigation District, 33–34; land redistribution law (1926), 74; mapped Colorado River delta, 25; increasing water use, 153; official unions, 133; opposed river diversion, 18; opposed foreign landowners, 74; policy towards its repatriates, 81; prohibited foreign landownership near borders, 13; protested Imperial Valley repression, 137; regulation of Compañia de Terrenos y Aguas de la Baja California, S.A., 30; Secretary of Communications and Public Works, report, 153; set water rates, 32; treaty with China, 106; unions in US, 133–134; Water Commission, 22; water rights, Colorado River, 8, 166; xenophobic riots, 100
Meza, Sarah, 163
military plans seize Alamo Canal, 131
militia, Mexicali, 131
mining, Baja California, 102
Minute 319, 173
Molino, Joe, 132
Moore, Chester: secretary, Western Growers and Shippers Protective Association, 92; manager, Associated Labor Bureau, 111; opposed Mexico quota, 116, 138–139
morality, Progressive Era ordinances, 54; enforcement of, 161
Mother Earth, 1–2
Muñoz, Germán, 172
mutual water companies, 16; debt, 31; operations, 43–44; purchased by Imperial Irrigation District, 32; water stock, 41
mutualista, El Centro, 133

Nation, civil liberties violations, 150
National Chamber of Agriculture, Mexicali, labor recruiter, 104
National Industrial Recovery Act, 142
National Industrial Relations Act, 123
National Irrigation Congress, 25
National Labor Relations Act, 156
nativists, Mexicali, opposed Japanese colonization, 34
nativists, US, opposed Japanese colonization in Mexicali Valley, 33–34
nature, engineered, 3
New Deal, 120–123, 126; supported agribusiness, 150
New Liverpool Salt Company, 21

New River: drainage, 43–44; pollution, 164
Nice, Charles, 149
Nimura, Takanori, 55
Nimura, Tanizo, 69–70
Northern District of Baja California, 12

Obregón, Alvaro, 30, 53, 74
Ockerson, John, 26–27
Okies, 125–126; culture, 118; Democratic Party, 123; labor surplus, 115; racialized, 115, 126
Olson, Culbert, 123–124
One-Man Farm program, 51
O'Neil, J. E., 105
Open Forum, complaint against fascism in Imperial Valley, 151
Oriental Exclusion League, support for Alien Land Law (1920), 84
Ortiz Rubio, Pascual, Mexicanizing Baja California, 153–154
Osborne, Hugh, 92–93, 127
Otis, Harrison Gray, 25, 43
Oxnard, 100

Palmerlee, Reverend M., 1
Panama Canal, comparison to flood, 22
Passbook Plan, 110–111
pests, eradicating, 61–66; permits to extinguish, 168
Phelan, James D., 68, 84
picketing caravans, 147–148
Pinchot, Gifford, 32
pioneers, 1, 3
Pioneers Museum, 48
Pisani, Donald, 47
poison, 6, 8, 62–63, 164
police brutality, 141, 144. *See also* repression
policy recommendations, author, 174–176
pollution, 163–164, 168
Poiriez, Brad, 171
Pomeroy, Harold, 122
poverty, 118, 126, 169, contemporary, 163–164
present perfected rights, water, 166
produce, 9
prostitution, 54–55

Quantification Settlement Agreement (QSA) 170–172
quarantine, 94; pests, 62

race relations, 138; alliances, 54; fluidity, 55; friendships, 56; sexuality, 78; Southerners, 55–56
racial: agriculture, 68; borderlands, 68–69; discord, 138; privileges, 121; stereotypes, 107–108
racism, 101, 138
railroad cars, statistics, 106
Ramos, Joe, 81
Reagan, Ronald, 48
Reclamation Act (1902), 14, 166; enforcement of, 34–35; ideology, 17–18
reclamation, ideology, conquest of nature, 14, 41
reclamation movement, 12
Reclamation Service, 11; All-American Canal preliminary report, 32; survey route, 31; alliance with Imperial Irrigation District, build aqueduct, 12, 29; discredited colonization, 17–19; negotiation to buy California Development Company, 46; 160-acre limitation, 29; Yuma project, 19, 25, 29
Reconstruction Finance Corporation, 120; poison, 63
red baiting, 136, 139, 141, 145, 147; during strike, 132
Red Cross, 118–119
red scare: Imperial Valley, 130; Mexicali, 131
Regional Labor Board, 144–146
relief (welfare), California, 119
relief (welfare), Imperial Valley: disparities, 120; during strikes, 120–121, 147; excluded aliens, 120
relief (welfare) Mexicali Valley, 125
relief (welfare), Imperial-Mexicali Valley, 118
remittances, 138
repatriation, to Mexico, 116, 125, 140, 153–154
repression (labor, political, social), Imperial Valley, 28, 53–54, 127–128, 134–142, 144–153, 157, 160–161, 169; ideological roots, 41; of Japanese, World War II, 167; origin, against labor, 132
repression, Mexicali Valley, 132, 140; during strike, 131
Republican: agribusiness elite, 54; political party, 41
resistance strategies, 6, 8–9, 54, 57–58; labor, 127–128. *See also* strikes
Rockwood, Charles, 12; closing river break, 21–22; floods, 19; land speculator, 42; railroad loan, 21; replaced George Chaffey, 17; resigned, 22
Rodríguez, Abelardo, 131–132; met consul Joaquín Terrazas in Mexicali, 152; ousted governor Esteban Cantú, 53
Roosevelt, Franklin, 21, 24, 120, 148, 156; All-American Canal funding, 36; strike telegram, 146
Rose, Mark: All-American Canal proposal, 29; elected to build aqueduct, 31; establishing mutual water companies, 46; land speculator, 46; atop excavator, 37; promoted aqueduct, 46
Rugg, Fred, 114

Sacramento Bee, 168
salinity crisis, protest outside US consulate, Mexicali, 172
Salvation Army, 119
San Diego County, 46
San Francisco, 100; lenders to Chinese tenants, Mexicali Valley, 74
San Joaquin Valley: cooperative, 81; strike, 120–121, 142
Sanbonmatsu, Teru, 78–80
Sanbonmatsu, Yoshimitsu, 79–80
sand dunes, 4, 13
Sanga, A. G., 82
Sano, Ichizo, 69
Saund, D. S., 86
Salton Sea, 50; biological collapse, 170, 174; biological mitigation, 171; creation, 11–12, 22, 25; flooding, 59–60; homesteading near, 50; hypersalinity, 175, pollution, 164; poisons, 170–171; salinity, 60, 170
secretary of interior, 11; international boundary survey, 49; opposed colonization, 18, 21; water master, Colorado River, 166, 170
segregation, racial, 9, 51, 54–56; resisted, 56
sexual competition, interracial, 109
shed. *See* labor, shed
Sherman, Moses, 43
Shields, Tina, 174
Shigematsu, M., 84–85
shippers, 68; control overproduction, 93–94; packinghouses, 73
Sinclair, Upton, 121–122
Sinophobia, Mexico, 74
Smith and Rogers, colonization company, 86
Smythe, William, 14; opposed California Development Company, 19; reclamation advocate, 47
social reforms, undermined, 118
Social Security Act, 123, 156
Sociedad de Irragación y Terrenos de la Baja California, S.A., 13, 23; regulated by Mexico, 18, 20; managed by Southern Pacific Railroad, 21; Mexican concession, convey water, 20, 38
socio-economic justice, farm workers, 166
Soil Conservation Committee, 96
Sonora, Mexico, 13; Sonoran dynasty, 74
South Asians (Hindus): farmers, 76–78; homosexuality, 77; Punjabi culture, 76–77
Southern Pacific Railroad, 11; California Development Company loan request, 21; colonization assistance, 42; flood prevention, 38; flood reimbursement, 28; lawsuit to return stock, 26; managed California Development Company, 48; negotiation to sell California Development Company, 29–30; reduced fares for laborers, 102; selling water stock, 45; in Yuma, Arizona, 17
speculators, land, 42
spies, 132, 141
squatters: Imperial Valley, 50; Mexicali Valley, 131–132
Standlee, Mrs., 89
starvation, Brawley, 119; Mexicali, protest march, 118–119
State Emergency Relief Administration (SERA), 120, 122–123
steamships, transport labor, from Sonora and Sinaloa, 102, 104, 106
Steinbeck, John, 124
strikes, California, 142
strikes, field labor, Imperial-Valley, 10; (1922), 131–133, 157; (1928), 133–139; (1930), 139–142; (1934), 142–153; (1960–1961), 169; agribusiness committee report, 151; Calexico meeting, 150; communist organizers, 148; conciliator, 150, 156; deportation, 135, 138; election for bargaining agent, 151–152; female leadership, 144; interracial, 157; labor conditions, hearings, 151; leaflets, 135; murder (1979), 169; $1,000 bail, 135; overview, 128–129; pea, 121. *See also* Pelham Glassford; picketers, ethnic and racial groups, 148; picketing, 144; squatter camps, 147–148; squatter camps burned, 148; scabs, 132, 147, 156; stockade, 145; telegram to Mexican government, 134–135; wildcat, 128
strikes, shed, 10, 127, 140–141, 144, 155–156, 161; death picketers, 156, violence, 156; wildcat, 128
structure of power, 45
subsidies, 39, 41, 60, 66–67, 69, 93–97; benefitted large operators, 89; water, 165; water transfers, Imperial Irrigation District, 171–172
Supreme Court, California, Alien Land Law (1920), 85–86; upheld legality of Imperial Irrigation District, 28
Supreme Court, US: denied Asian naturalization, 69, 85; *Arizona v. California* (1963), 165–166; *Arizona v. California,* supplemental decree, 166, 174; Ben Yellen, lawsuit (1980), 166, 171, 174; Industrial Workers of the World, 130; legacy of rulings, 166–167
Suarez-Potts, William, 133
Swing, Phil, 31; All-American Canal legislation, 32; lobbying for aqueduct funding, 36; 160-acre exemption, 36; opposed Japanese colonization, Mexicali Valley, 87

Taft, Clinton, 141
Taft, William Howard, 26

tarred and feathered, 7, 21
Taylor, Paul, 137; Imperial Valley research, 81; migratory labor studies, 117–118
tenants, Asian, 68; credit, Imperial Valley, 71–72; census (1910), 70; census statistics, 83, 88; contemporary, 167; living conditions, 82–83; Japanese, 75–76; organization, 82–83; racial, 82–83; racial animosity, 83; Southern, 68; stereotypes, 71–73; white, 71–72
tenants, Mexicali Valley, Asian, 68
Terrazas, Joaquín, 128, 143–145, 147–153, 158; met president Abelardo Rodríguez, Mexicali, 152; threats against, 149
Texas, tenants and pickers, 75
Thurman, Roscoe, 130
Tijuana, 132; magonista invasion, 27–28
Tokeshi, Doso (and wife), 78
Torres, Hermolao, 134–138
Torres-Martínez reservation, 96; flooding, 59–60
Tout, Otis, 23–24, 45
Trade Union Unity League (TUUL), 139, 141

Unión de Trabajadores del Valle Imperial, 133–134, 143–144
United Farm Workers, 169
US, government relations with settlers, 49–50
US-Mexico borderlands, ix; boundary survey, incorrect, 17, 49; instability, 131–132
US-Mexico relations: antagonistic, 29; labor recruitment, 105; Mexican transnational migration, 98, 160; water treaty (1944), 172–173
University of California, Berkeley: pest control, 62; subsidies, 93

vagrancy, 133, 135; law, 56, 129; vagrants, 57
Valdés, Luis, 148
Van Dimas, Paul, 21
Vásquez, Gambino, 155
Vega, Celso, 27–28
vegetables, 1920s, 75–76
Velásquez viuda de Arellano, Felipa, 140
veterans, World War I: homesteaders, 50; One-Man Farm program, 51; sentinels, 51; supported All-American Canal, 35
vigilantism: legacy, 157; vigilantes, 146, 149–152
Villa, Francisco "Pancho," 131
voters, Imperial County: approved Imperial Irrigation District policy and contracts, 9, 31–32, 35; public ownership of water utility, 38; minorities lacked suffrage, 54

wages, Imperial Valley, 112, 168; non-strike bonus (holdback), 113, 138, 156; claims, 112–113; during Great Depression, 115–118; for Mexicans, 108
wages, Mexico, 102
Wagner, Madge Morris, 40
Wagner Act, 123
Waldron, Francis, 139
Warren, Earl, 79
water: Imperial Irrigation District-San Diego water transfer, 170; scarcity, 163; shaped land, 59; stock, 16; rural to urban transfers, 169–172, 174; socio-economic inequality, 172; US-Mexico competition over, 59; waste, 163, 169–170
Water Commission, Mexico, 22
water conservation, Imperial Irrigation District, 170–172
water distribution, 45
water, reasonable and beneficial use, 175
water rights, Colorado River: California apportioned, 35; California Development Company notices, 14, 30; California Development Company/Imperial Irrigation District claim, 38; Imperial Irrigation District claim, 30, 166; Imperial Irrigation District dubious claim, 165–166; inflated, 12; US, 38, 166; US and Mexico, 8; study, 34
Webb, US, 86
Western Growers and Shippers Protective Association, 92, 116, 139, 143
Westmorland, 86; rabbit hunt, 65
Westmorland, Roy, 70
Wheatland, 129
white spot, 117
white supremacy, reclamation movement, 166–167
Wiest, G. E., 64
Williams, Chester, 151
Winning of Barbara Worth, 41, 47–48
Wirin, A.L., 146
Witter, Myron, 45, 84
wobblies. *See also* Industrial Workers of the World; magonista invasion of Mexicali, 27, 129–133
Wonders of the Colorado Desert, 1
workers. *See* labor
Works Progress Administration, 120, 123
World War II, 168
Wrap Pack Association, 156
Wrestled with the River, 3
Wright, Harold Bell, 41, 47–48

xenophobia, Mexicali Valley, 131, 153–154, 160

Yaqui army, 131
Yellen, Ben: lawsuit enforce 160-acre limitation, 165–166, 169; legacy, 174

Zinn, Evelyn, 63

www.ingramcontent.com/pod-product-compliance
Lightning Source LLC
LaVergne TN
LVHW041948310525
812621LV00004B/13

* 9 7 8 1 6 2 3 4 9 4 6 3 6 *